Introduction to Probability Theory with Computing

Introduction to Probability Theory with Computing

J. Laurie Snell

Dartmouth College

Prentice-Hall, Inc., Englewood Cliffs, New Jersey

Current printing (last digit):

10 9 8 7 6 5 4 3 2 1

Printed in the United States of America
ISBN: 0-13-493445-8

Prentice-Hall International, Inc., London
Prentice-Hall of Australia, Pty. Ltd., Sydney
Prentice-Hall of Canada, Ltd., Toronto
Prentice-Hall of India Private Limited, New Delhi
Prentice-Hall of Japan, Inc., Tokyo

CONTENTS

I
PROBABILITY MEASURES

II
RANDOM VARIABLES

III
LIMIT THEOREMS

IV

MARKOV CHAINS

PREFACE

This book is an introduction to probability with computing. The probability theory is limited to the case of experiments with a finite number of outcomes. Limit theorems are discussed but only as an approximation to the finite case.

Chapters 1,2, and the first three sections of Chapter 3 are the core material of an introduction to finite probability. This includes sample spaces, probability measures, random variables, sums of independent random variables,the law of large numbers and the central limit theorem.

Sections 4 and 5 of Chapter 3 discuss fluctuation theory following the elegent treatment presented by Feller in his classic book "Probability Theory and its Applications". The material in these sections is more difficult than that in the rest of the book but it is a good topic to illustrate the use of simulation.

Chapter 4 is devoted to Finite Markov Chains. The treatment is basically that presented in "Finite Mathematical Structures", which the author wrote with J.G.Kemeny, H. Mirkil and G. Thompson. Appropriate computing has been added. The author was a coauthor of this book and is grateful to John G. Kemeny, Gerald L. Thompson and to Prentice-Hall for permission to make use of Chapter 6 of their book, "Finite Mathematical Structures", (C) 1959, Prentice-Hall Inc., Englewood Cliffs, New Jersey.

The programming language used is BASIC and the honest reason is that this is the only computing language the author knows.

The core material can be read without calculus but an

introduction to calculus would be helpful. A knowledge of matrix operations is assumed in Chapter 4. Of course, the availability of a computer and the knowledge of BASIC is assumed.

This book grew out of a course given jointly with Professor John G. Kemeny on computing and probability. The author is indebted to Professor Kemeny for convincing him some years ago that finite mathematics was fun and the same is true of computing. Any originality in the book is the result of the authors years of joint teaching and research with President Kemeny. Were he not engaged in trying to keep Dartmouth College afloat, it might well have been a joint book--better serving justice to say nothing of the spelling. However, the author will have to accept the blame for any errors in the book.

Professor Skip Weed read the manuscript with great care and made a number valuable suggestions. All were taken except one or two obviously designed just to make it easier for him to teach his statistics courses. David Griffeath helped make up problems. If you find a problem too difficult it was probably contributed by David. David and Bob Beck introduced me to the possiblities of computer graphics and wrote the programs which provide the computer graphics in the text. Ruth Bogart gave valuable editorial advice as well as programming assistance.

A secondary aim in writing this book was to see if it is practical to write a mathematical text on a teletype and to have the text on tape for editing and future modifications. David Thron was a great help in proving this to be possible. So was Jane Goldberg who was willing to type into a computer and put up with the computers responses such as "Jane is not saved". Since the text was photographed from a computer printout the author was limited in the number of symbols that could be used and the justification of lines is a bit crude. However this procedure does make the text easy to modify the author would appreciate suggestions for improvement on this preliminary version.

Finally, there is the computer itself and the Dartmouth computing center. Obviously, the whole project would not have been possible without them. And there is probably no computing center that is more helpful and available than Dartmouth's. However, it will take some time for the frustrations of dealing with a computer to wear off before the author feels that he can give, with honesty, the thanks that are certainly due.

J.L.S.

I

PROBABILITY MEASURES

1. INTRODUCTION

While the mathematical theory of probability is relatively new, the problems that it deals with are very old. One of the first classes of problems studied was the class of problems which arose in gambling games. For example, one might be asked to compare the chance of obtaining a 2 with that of obtaining a 7 when a pair of dice are thrown. A 2 can result only if both dice turn up a 1 and a 7 can occur in several different ways. Well before any formal developement of probability it was realized that the chance of a 7 is greater than a 2.

Although archeologists have found evidence of dice-shaped bones in prehistoric sites, the formal study of these problems is generally felt to have started in the sixteenth century. In particular one of the first books on the subject of games of chance was the book "Liber de Ludo Aleae", "The Book on Games of Chance" written by an exceptionally colorful man named Gerolamo Cardano. Cardano was a man of many trades including medicine, science, and astrology. An interesting account of the life and work of Cardano may be found in the book "Cardano, the Gambling Scholar" by O. Ore.

The progress of probability continued to be served by gamblers asking men of science to explain their experiences in games of chance. For example, Galileo was asked why the total 10 appears more often than 9 in throws of 3 dice. Pascal was asked why it is more likely to obtain a 6 in four rolls of a die then it is to obtaintwo 6's in 24 rolls of two dice. This question led to a famous series of letters between the mathematicians Pascal and Fermat. They were led to study a problem called the *problem of points*. In this problem two players of equal skill are playing a game in which the players make a series of plays each play resulting in a point for one of the players. The first player to win N points is declared the winner. The problem that they solved was that of deciding how to divide the prize if the players are forced to quit before the end of the game with one player having A points and the other B. We shall study these problems and many others. From this modest beginning a theory of probability evolved and applications have been made to almost every field of study--be it in science, social science, or in the humanities. While the scope of the theory and applications has increased immensely, the methods developed by the early works of Fermat, Pascal, and others in the sixteenth and seventeenth centiries are still an important part of modern techniques for solving problems in probability theory.

Before we begin our formal study of probability it is interesting to consider our intuitive ideas of probability. For example, consider the familar experiment of tossing a coin--an experiment usually carried out at the beginning of a football game to decide which team will kick off. Our intuitive idea of this experiment is that it results in one of two outcomes, heads or tails, and does not favor either outcome. A little reflection shows that there are difficulties in realizing even this simple experiment. For example, we would want the two sides of the coin to be exactly the same so that neither side would be favored. Also we would want to be sure that the coin is spun in such a way that the knowledge of the face turned up before spinning the coin does not influence the final outcome. One might even argue that if we knew the forces on the coin when it is tossed we should be able to predict, by the laws of physics, the exact trajectory of the coin and hence its final outcome--leaving nothing at all to chance. Even if we were happy with the experiment we have to say what we mean by the probability that the outcome is heads. Most people would say that the probability is 1/2. Some because of the symmetry of the coin and others because they believe that if the experiment were repeated a large number of times heads would turn up about half the time. Early attempts at formulating probability theory were along the lines of this frequency concept. However, trying to make this precise leads to difficulties. For example, we do not expect to get a head exactly half of the time. It is remotely possible that heads could turn up every time in 1000 tosses. All we can say is that it is very "likely" that the proportion of heads will be near 1/2. But what do we mean by likely? An answer such as "something that will occur with high probability" only leads in a circle.

Similar difficulties were encountered in the developement of plane geometry. It is easy to imagine drawing a straight line. However, we would really like it to have no width and indefinite length. Euclid wanted to give defintions for points and lines. In modern treatments of his plane geometry, points and lines are undefined objects satisfying certain axioms. Theorems are then derived from the axioms by rules of logic. In the course of proving theorems one can draw pictures and use intuitive ideas about points and lines to suggest theorems and their proofs. The theorems in turn give insights into practical problems which are approximated by the geometry.

Like geometry, modern probability is based on an axiomatic system which was suggested by certain intuitive concepts of probabilities. Within this framework, a basic theorem, the *law of large numbers*, gives a meaning to the frequency concept. The theory that we develop helps to explain physical experiments involving chance outcomes, and our intuition about such experiments is useful in suggesting further results.

The axiomatic theory of probability theory is particulary simple for the case of experiments which have only a finite number of possible outcomes. Most of our discussion will be limited to this situation. Fortunately, though the mathematics is relatively simple for this case there is still a wealth of applications that one can make using only the theory of finite experiments.

We have only touched upon the historical background and the difficulties met in defining probability. We shall turn to the mathematical developement next. The reader who wishes a more extended discussion of these problems is invited to read the delightful "Letters on Probability" by A. Renyi. Books which give a detailed developement of probability theory are "Games, Gods, and Gambling" by F. N. David and "Probability Theory, a Historical Sketch" by L. E. Maistrov.

2. SAMPLE SPACE

The first step in studying a chance experiment is to describe the possible outcomes of the experiment. If we roll a die we may describe the outcome as one element of the set $\Omega = \{1,2,3,4,5,6\}$ of possible outcomes. If we toss a coin twice we record the outcomes as $\Omega = \{HH,HT,TH,TT\}$. In each case, the set is called the <u>sample space</u>.

We require of our description that one and only one outcome must occur. There is, within this requirement, still some choice depending on how detailed a description we find useful. For example we may be interested only in the number of heads which turn up when a coin is tossed twice. Then we could choose $\Omega = \{0,1,2\}$.

Consider then an experiment with sample space $\Omega = \{a,b,c,\ldots,r\}$. When we wish to refer to the outcome of the experiment without specifying the actual value of the outcome we shall use the notation ω. An <u>event</u> is a subset E of the sample space Ω. We say that the event E "occurs" if the outcome is an element of E.

<u>Example 1</u>. A coin is tossed twice. We take as sample space $\Omega = \{HH, HT, TH, TT\}$. The event E, "heads on the first toss," is the subset $E = \{HH, HT\}$. The event F, "at least one head turns up," is the subset $F = \{HH, HT, TH\}$.

Our aim is to determine the probability $P(E)$ that the event E occurs. We define this for all events relative to Ω as follows. We assign to each element of Ω a non-negative number $P(\omega)$. We assume that

$$(1) \quad \sum_{\omega} P(\omega) = 1$$

This corresponds to the fact that one of our elementary events ω which describe the sample space must occur. If E is any event, we define P(E) to be

$$(2) \quad P(E) = \sum_{\omega \text{ in } E} P(\omega)$$

That is, P(E) is the sum of the probabilities for all outcomes ω which cause the event E to occur.

Example 1 (Cont.). Let us assign equal probabilities to each of the four possible outcomes for the two tosses of a coin. That is, $P(HH) = P(HT) = P(TH) = P(TT) = 1/4$. Then the event E, "heads on the first toss" has probability

$$P(E) = P(HH) + P(HT) = 1/2$$

and the event F, "at least one head turns up" has probability

$$P(F) = P(HH) + P(HT) + P(TH) = 3/4.$$

Example 2. Three men A,B, and C are running for the same office. We feel that A and B have the same chance but C has only 1/2 the chance of A or B of winning. Adopting an obvious notation, this suggests that we assign $P(A) = P(B) = 2P(C)$. Since

$$P(A) + P(B) + P(C) = 1$$

we see that we must have

$$2P(C) + 2P(C) + P(C) = 1$$

or $P(C) = 1/5$. Then $P(A) = P(B) = 2/5$. Let E be the event "A or C wins." Then $E = \{A,C\}$ and $P(E) = 1/5 + 2/5 = 3/5$.

We return now to the general case. We note that by (1) and and (2) we assign to each subset E of Ω a number $P(E)$. We may think of $P(E)$ as defining a numerical valued function with domain the set of all subsets of the sample space Ω. We shall call $P(E)$ a _probability measure_. In order to describe some of the properties of this measure we shall need certain standard operations on sets.

DEFINITION. If A and B are two subsets of Ω, the _union_ $U = A + B$ of these sets is the set U consisting of all elements of Ω which are in either A or B. The _intersection_ $V = AB$ is the set of all elements of Ω which are in both A and B. The _complement_ of a set A, A' is the set of all elements of Ω which are not in A.

An alternative notation for the union $U = A+B$ that is often used is $U = A \cup B$. Similarly, the notation $V = A \cap B$ is often used for the intersection $V = AB$. Finally A' is often denoted by $\tilde{A}$.

We shall denote by $\emptyset$ the empty subset of Ω. That is, the subset consisting of no elements. This would correspond to an event which could not occur and $P(\emptyset) = 0$.

The diagrams in Figure 1 called <u>Venn diagrams</u> are useful in visualizing these set operations. The shaded areas represent the subsets AB, A+B and A' respectively.

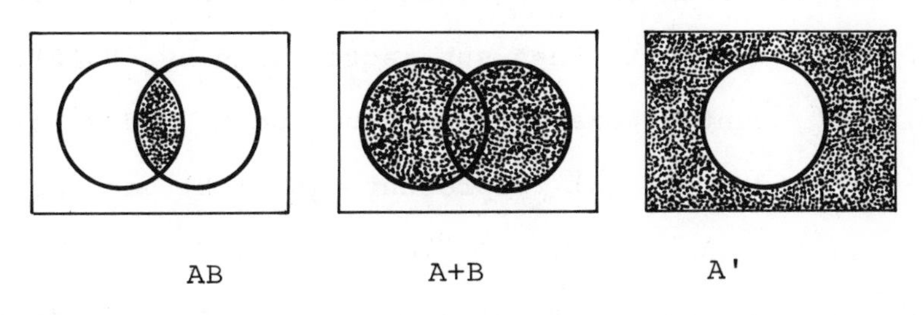

AB A+B A'

Figure 1.

We can now state some basic properties of a probability measure in terms of these set operations.

THEOREM. Assume that $P(E)$ is a probability measure defined on a sample space Ω. Then if A and B are any subsets of Ω:

(a) $0 \leq P(A) \leq 1$

(b) $P(A + B) = P(A) + P(B) - P(AB)$

(c) $P(A + B) = P(A) + P(B)$ if $AB = \emptyset$

(d) $P(A') = 1 - P(A)$.

Proof. To prove (a) we need only remember that $P(A)$ is the sum of certain of the non-negative numbers $P(\omega)$. The sum of all the $P(\omega)$ is 1. Thus $0 \le P(E) \le 1$.

We prove next (b). $P(A + B)$ is the sum of all $P(\omega)$ for ω in either A or B. Assume first that ω is in one but not both of these sets, say A. Then on the right side of (b) $P(\omega)$ is counted once by $P(A)$ and not by either of the terms $P(B)$ or $P(AB)$. Thus it is counted once by both sides of (b). Assume next that ω is in both A and B. Then on the right side of (b) it is counted twice by $P(A) + P(B)$ but then subtracted once by $P(AB)$. Thus again $P(\omega)$ is counted exactly once by both sides of (b). Therefore both sides of (b) represent $P(A + B)$ as was to be shown.

Property (c) follows from (b) and the fact that $P(AB) = 0$ if $AB = \emptyset$.

To prove (d) we note that

$$\Omega = A + A'$$

and by (c) $P(\Omega) = P(A) + P(A') = 1$.

Example 3. Let us illustrate these properties in terms of three tosses of a coin. When we have an experiment which takes place in stages such as this, we often find it convenient to represent the outcomes by a tree diagram as shown in Figure 2.

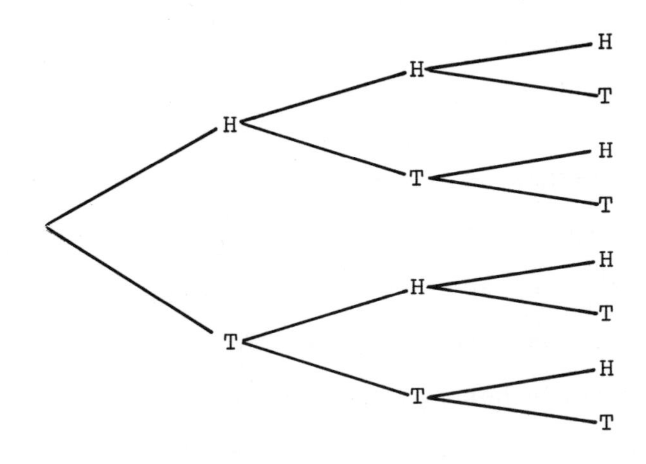

Figure 2.

A <u>path</u> through the tree represents a possible outcome for sequence of individual stages of the experiment. For the case of three tosses of a coin we have eight paths and again we assign an equal weight, 1/8, to each path. Let E be the event "at least one head turns up." Then E' is the event "no head turns up." This event can occur in only one case, namely ω = (TTT). Thus P(E')= 1/8 and by property (d) P(E) = 1-1/8 = 7/8. We shall find that it is often easier to compute the probability that an event does not happen than that it does. We then use property (d) to obtain the desired probability.

Let A be the event "the first outcome is a head." Let B be the event "all three outcomes are tails." Then P(A) = 1/2 and P(B) = 1/8. Since AB = $\emptyset$, we have by (c) P(A+B) = P(A) + P(B). Thus P(A+B) = 1/2 + 1/8 = 5/8. Since

$$A + B = \{HHH, HHT, HTH, HTT, TTT\}$$

we see that we obtain the same result by direct enumeration.

In our coin tossing example we were led to assign an equal probability to each possible outcome. This method of assigning a probability measure is given a special name, the equiprobable measure. If this measure is assigned to a sample space with n elements, each possible outcome has probability $1/n$ of occuring. The probability of an event E with m elements is m/n. Thus for the case of the equiprobable measure, calculating probabilities of an event reduces to a counting problem. We shall be able to discuss more significant examples after we have reviewed some counting techniques.

EXERCISES

1. A student must choose one of the subjects art, geology or pyschology as an elective. She is equally likely to choose art or psychology, and twice as likely to choose geology. What are the respective probabilities that she chooses art, geology, and psychology?
2. Another student must choose exactly two out of three electives: art, French, and mathematics. He chooses art with probability 5/8, French with probability 5/8, and art and French together with probability 1/4. What is the probability that he chooses mathematics? What is the probability that he chooses either art or French?

Ans. 3/4,1.

3. Describe in words the events described by the following subsets of

   Ω = {HHH, HHT, HTH, HTT, THH, THT, TTH, TTT}.

   a. E = {HHH, HHT, HTH, HTT}

   b. E = {HHH, TTT}

   c. E = {HHT, HTH, THH}

   d. E = {HHT, HTH, HTT, THH, THT, TTH, TTT}

4. What are the probabilities of the events described in Exercise 3?

5. A die is rolled once. The sample space is

$$\Omega = \{1,2,3,4,5,6\}.$$

   What is the probability that:

   a. The outcome is an even number?

   b. The outcome is less than 5?

   c. The outcome is a 2 or a 3?

   d. The outcome is greater than 7?

6. A and B are events such that A is a subset of B. That is, every element of A is also in B. Show that $P(A) \leq P(B)$.

7. John and Mary are taking a mathematics course. The course has only three grades, A, B, and C. A is the highest grade and C the lowest. The probability that John gets a B is .3. The probability that Mary gets a B is .4. The probability that neither gets an A but at least one gets a B

B is .1. What is the probability that at least one gets a B but neither gets a C?

Ans. .6

8. If A, B, and C are any three events show that:

$$P(A+B+C) = P(A)+P(B)+P(C)-P(AB)-P(AC)-P(BC)+P(ABC)$$

9. A card is drawn at random from a deck of playing cards. Let E be the event "The card is an honor card," (i.e., an ace,king,queen,jack,or ten), let F be the event "The card is a heart," and G the event "The card is either a spade or a king". Find $P(E+F+G)$.

10. A die is loaded in such a way that the probability of each face turning up is proportional to the number of dots on that face. (For example, a 6 is three times as probable as a 2.) What is the probability of getting an even number in one throw?

Ans. 4/7

11. Two die are thrown. Describe a suitable sample space for this experiment and find:

    (a) The proability that the sum is seven.

    (b) The proability that the maximum outcome is a 3.

3. TWO COMPUTER APPLICATIONS

We have seen in the previous section that a chance experiment with a finite number of outcomes $\omega_1, \omega_2, \ldots, \omega_r$ will be described mathematically by assigning a probability measure to the space Ω of possible outcomes. We do this by assigning to each ω_j a non-negative number $P(\omega_j)$ in such a way that

$$P(\omega_1) + P(\omega_2) + \ldots + P(\omega_r) = 1$$

To illustrate our probability theory it will be convenient to be be able to perform an experiment which corresponds to given probabilities. For example, assume that $P(\omega_1) = 1/2$, $P(\omega_2) = 1/4$ and $P(\omega_3) = 1/4$. Then we could realize these probabilities by the following experiment. Toss two coins. If they turn up with different faces call the outcome ω_1. If both are heads call it ω_2. If both are tails call it ω_3. Then $P(\omega_1) = 1/2$, $P(\omega_2) = 1/4$ and $P(\omega_3) = 1/4$.

It would be difficult if for every choice of probabilities we had to think up a different kind of experiment. Fortunately, we can describe one experiment that will work for any set of probabilities. It has the difficulty that the possible outcomes are not finite, but we have already warned the reader that we shall have to delve into the infinite at times.

Consider a circular wheel with unit circumference and with a pointer as shown in Figure 3.

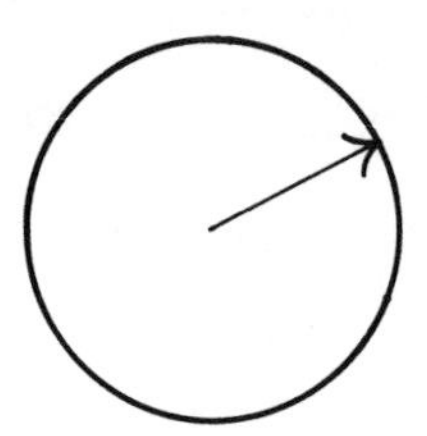

Figure 3.

Suppose we want to realize our experiment with $P(\omega_1) = 1/2$, $P(\omega_2) = 1/4$ and $P(\omega_3) = 1/4$. To do this we mark off three intervals I_1, I_2, I_3 of length 1/2, 1/4, 1/4 respectively as shown in Figure 4.

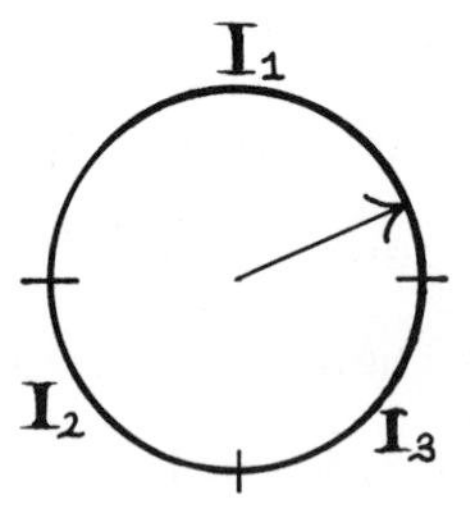

Figure 4.

We spin the pointer. We say that ω_1 occurs if the outcome is in interval I_1, ω_2 if it is in I_2, and ω_3 if it is in I_3. It would seem reasonable to assume that the pointer is twice as likely to stop in I_1 as I_2 and have the same chance of landing in the equal intervals I_2 and I_3. That is, $P(\omega_1) = 1/2$, $P(\omega_2) = 1/4$, $P(\omega_3) = 1/4$. Thus we have a model for our given probability measure. The advantage of this model over our previous coin tossing model is that it can be modified to serve as a model for any finite probability measure.

For the general case $P(\omega_1), P(\omega_2), \ldots, P(\omega_r)$ we proceed in the same way. We mark off intervals $I_1, I_2, \ldots, I_r$ with I_j having length $P(\omega_j)$. Since the circle has unit circumference and

$$\sum_j P(\omega_j) = 1$$

this is possible. If E_j is the event "the pointer stops in interval I_j", it would again seem appropriate to assign a probability to this event equal to the length of the interval.

As we have mentioned we shall be particularly interested in seeing what happens when we repeat a chance experiment a large number of times. While our spinner is a convenient way to carry out a few repetitions it would be rather difficult to carry out a very large number. Since the computer can do operations extremely fast it is natural to try to find a way to carry out this experiment by the computer. We first consider the analog of a single spin of a wheel. This is done by means of a random number generator. This is a process by which the computer attempts to pick a number between 0 and 1 in such a way that the probability that the number lies in any particular subinterval of the unit interval is equal to the length of this subinterval. For our basic example $P(\omega_1) = 1/2, P(\omega_2) = 1/4, P(\omega_3) = 1/4$ we would divide the interval as in Figure 5. We say that ω_1 occurs if the random number produced lies in I_1, ω_2 occurs if it lies in I_2 and ω_3 if it lies in I_3.

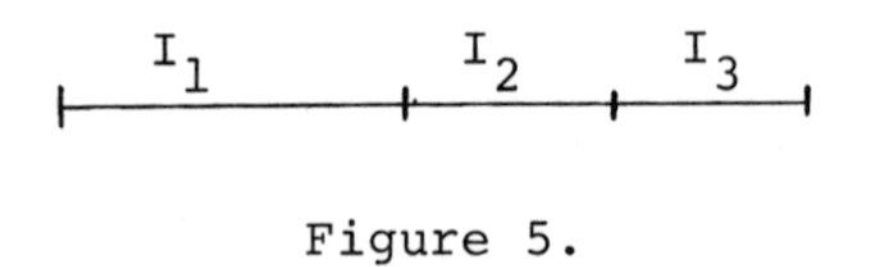

Figure 5.

It would take us too far afield to discuss the method by which random numbers or "pseudo random numbers" are generated. However, to obtain from the computer a random number is quite easy. In BASIC one simply asks for RND. Every time you ask you get a different random number between 0 and 1. The program RANDOM chooses a sequence of 20 random numbers between 0 and 1. We note that our computer only prints out numbers to six decimal places.

```
RANDOM
10 RANDOMIZE
20 FOR I = 1 TO 20
30    PRINT RND,
40 NEXT I
50 END
```

Running this program yields twenty "random numbers."

```
RANDOM

.911985         .929211         .652637         .12811          .85835
.802956         .378397         .458021         .372047         .907573
.308569         .275879         .967226         .997203         .437282
.815806         .117997         .785947         .718628         .271883
```

If the instruction RANDOMIZE in line 10 of the program RANDOM were omitted, the program would produce the same outcomes each time it was run. While this is often desireable for debugging a program, we would want this instruction if we plan to make more than one run of the program.

We can modify this program slightly to toss a coin a sequence of times. As we have noted our intuition suggests that the probability of obtaining a head on a single coin toss is 1/2. Thus we should be able to "toss a coin" by the computer by testing each random number to see if it is less than 1/2. If so, we call the outcome "heads." If the number is greater than 1/2, we call it "tails."

```
COIN

10 RANDOMIZE
20 FOR I = 1 TO 20
30    IF RND < .5 THEN 60
40    PRINT "T";
50    GO TO 70
60    PRINT "H";
70 NEXT I
80 END
```

The program was run with the results

```
COIN

TTTTHHHTTTHHTHHHHTTT
```

Notice that in 20 tosses we obtained 9 heads and 11 tails. Let us toss a coin 10,000 times and see if we obtain a proportion of heads close to our intuitive guess of 1/2. We modify the program so that it no longer prints out "H" or "T" for each toss. Instead, we use a counter (line 50) to keep track of the number of heads.

```
COIN1

10 RANDOMIZE
20 LET S=0
30 FOR I=1 TO 10000
40    IF RND<.5 THEN 60
50    LET S=S+1
60 NEXT I
70 PRINT S/10000
80 END


COIN1

.4987


COIN1

.506
```

In each run of the program, we obtained a value close to 1/2.

The above method for estimating a probability by carrying out the experiment a large number of times on the computer is called <u>simulation</u>.

We can use the method of simulation to illustrate the experience of the gambler that queried Galileo. Recall that the question asked was whether it is more likely to obtain a 9 or a 10 as a sum when three dice are thrown. The program GALILEO carries out this experiment 50,000 times and prints the proportion of times that the sum is 9 and the proportion of times that it is 10.

```
GALILEO

100 RANDOMIZE
110 FOR I = 1 TO 50000
120    LET S = 0
130    FOR J = 1 TO 3
140       LET X = RND
150       FOR K = 1 TO 6
160          IF X < K/6 THEN 180
170       NEXT K
180       LET S = S+K
190    NEXT J
200    IF S = 9 THEN 230
210    IF S = 10 THEN 250
220    GO TO 260
230    LET U = U+1
240    GO TO 260
250    LET V = V+1
260 NEXT I
270 PRINT "PROPORTION OF  9's =";U/50000
280 PRINT
290 PRINT "PROPORTION OF 10's =";V/50000
300 END
```

The result of a run of this program is

```
GALILEO

PROPORTION OF  9's = .11606

PROPORTION OF 10's = .12824

GALILEO

PROPORTION OF  9's = .11682

PROPORTION OF 10's = .12636
```

The loop from 130 to 190 simulates three rolls of a die. To simulate the roll of a single die we choose a random number X and find the first integer K such that X $< K/6$. This number is the outcome of a single roll of the die. At the completion of the loop the sum is represented by the value of S. In lines 200 and 210 we test if this sum is 9 or 10. If it is 9 we increment U by 1 and if it is 10 we increment V by one. Thus after the loop 110 to 260 we have found the number of times in 50000 experiments of three rolls of a die, that a 9 occurs and the number of times that a 10 occurs. We print the proportions U/50000 and V/50000 as an estimate of the probability of obtaining a 9 or a 10 respectively.

From this experiment we might estimate the probability to be .117 for obtaining a 9 and .126 for obtaining a 10. By enumerating the 216 possible outcomes for the three throws Galileo showed that a 9 occurs in 25 cases and a 10 in 27 of the cases. Assuming that we assign the equiprobable measure the true probability for obtaining a 9 is 25/216 = .116 and for a 10 is 27/216 = .125. Thus our estimate by simulation is close to the true value.

It should be remarked that while we can get quite good estimates of probabilities by the method of simulation it is often at the expense of an excessive amount of computing time. Tossing a coin 10,000 times by COIN1 took only .5 second on the Dartmouth computer. However, the 50,000 experiments carried out by GALILEO took slightly more than 30 seconds. While this is not a long time it is expensive for the computer.

For our second application we return to Example 3 of the previous section, the experiment of tossing a coin three times.

We describe the outcomes again as a tree but this time we label an outcome heads by a 1 and tails by a 0. Also we label the paths 0,1,2,..,7. The resulting tree diagram is shown in Figure 6.

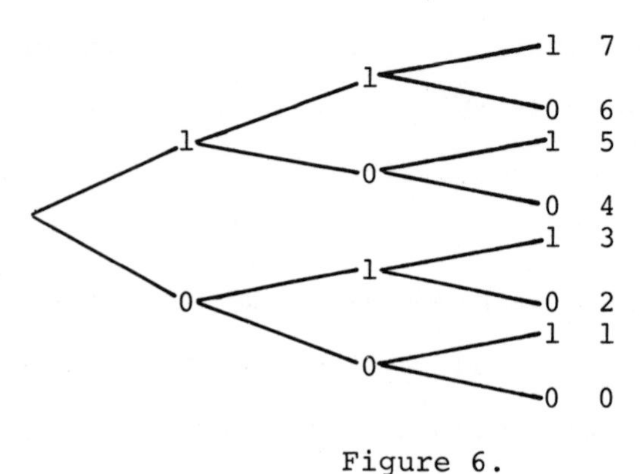

Figure 6.

To understand the correspondence between the path labels and sequence of 0's and 1's along the paths we must recall some facts about representing integers using the base 2 instead of the familiar base 10.

Every integer x between 0 and 2^n-1 can be uniquely represented in the form

$$x = a_{n-1}2^{n-1} + a_{n-1-2}2^{n-2} + \dots + a_1 2^1 + a_0 2^0$$

where $a_{n-1},\dots,a_1,a_0$ are all either 0 or 1. We write x then in the form $a_{n-1}\cdots\cdot a_1 a_0$ and call this a <u>binary representation of x</u>.

In particular the numbers between 0 and 7 can be written in the form

$$x = a_2 2^2 + a_1 2^1 + a_0 2^0$$

where a_2, a_1, a_0 are 0 or 1 and represent the binary digits for x. For example,

$$5 = 1x2^2 + 0x2^1 + 1x2^0$$

and so 5 would be written in binary representation as 101. Similarly,

$$3 = 0x2^2 + 1x2^1 + 1x2^0$$

so that 3 would be written in binary representation as 011. We see from Figure 6 that the digits in each case represent the outcomes along the path corresponding to the number being represented. The reader can check that this correspondence holds between each path and the binary representation for the path number. Note that to make this correspondence we must use all three digits even though the leading 0's do not affect the value of the number.

If we start with an integer x between 0 and 7 and want to obtain the binary representation we simply make 3 successive divisions by 2. The resulting remainders written in reverse order are the binary digits. For example, to find the binary representation of 5 we see that

5/2 = 2 + a remainder of 1

2/2 = 1 + a remainder of 0

1/2 = 0 + a remainder of 1.

The remainders 1,0,1 are The desired digits. Similarly if we start with 3

3/2 = 1 + a remainder of 1

1/2 = 0 + a remainder of 1

0/2 = 0 + a remainder of 0

giving the binary representation of 3 as 011. The simple "algorithm" illustrated above makes it easy to write a program to determine the binary representation of an integer. The program BINARY is such a program. The binary representation of each number A between 0 and $2^3-1 = 7$ is calculated in the loop from 150 to 250. This is accomplished as follows: In the loop from 150 to 200 the number A is divided by 2 three times. After the Jth division the remainder R is stored as the Jth component H(J) of the vector H (line 180). In lines 210 to 230 the components of H are printed in reverse order to give the binary representation of A. We note that a run of this program does produce the desired sequences representing the possible pathes through the tree.

```
BINARY

100 READ N
110 DATA 3
120 FOR A = 0 TO 2↑N-1
130    LET M = A
150    FOR J = 1 TO N
160       LET D = INT(M/2)
170       LET R = M-2*D
180       LET H(J) = R
190       LET M = D
200    NEXT J
210    FOR I = N TO 1 STEP -1
220       PRINT H(I);
230    NEXT I
240    PRINT
250 NEXT A
260 END
```

```
BINARY

0  0  0
0  0  1
0  1  0
0  1  1
1  0  0
1  0  1
1  1  0
1  1  1
```

Of course this would be a complicated way to obtain the paths in a tree for three tosses of a coin. But the same method applies for any number of tosses. Hence by just changing the value of N in the program we can generate all possible sequences of 0's and 1's of length N. To find the probability of an event relating to N tosses of a coin we need only put into our program a method to count the number of cases in which the event occurs. For example, we shall be interested in later sections in finding the probability that exactly J heads turn up when a coin is tossed N times. The program BINARY1 is a modification of BINARY to count the number of times exactly J heads turn up in 12 tosses of a coin for J = 0,1,2,...,12. We have only to introduce a counter S, line 200, which counts the number of 1's (heads) in the sequence. The vector H (line 230) keeps track of the number of times each particular value of S occurs as the program runs through all possible sequences of 12 tosses. In line 110 we have defined a function F(X) which is used to print a number to 4 decimal accuracy. It applies the usual convention of rounding up if the fifth digit is ≥ 5 and down when it is < 5. We shall often use this device to avoid complicated printouts.

```
BINARY1

100 DIM H(12)
110 DEF FNF(X) = INT(10000*X+ .5)/10000
120 READ N
130 DATA 12
140 FOR A = 0 TO 2↑N-1
150    LET M = A
160    LET S = 0
170    FOR J = 1 TO N
180       LET D = INT(M/2)
190       LET R = M-2*D
200       LET S = S + R
210       LET M = D
220    NEXT J
230    LET H(S) = H(S) + 1
240 NEXT A
250 FOR S = 0 TO N
260    PRINT S,H(S),FNF(H(S)/2↑N)
270 NEXT S
280 END
```

```
BINARY1

0              1              .0002
1              12             .0029
2              66             .0161
3              220            .0537
4              495            .1208
5              792            .1934
6              924            .2256
7              792            .1934
8              495            .1208
9              220            .0537
10             66             .0161
11             12             .0029
12             1              .0002
```

We note that the sequences which have 6 heads occur more often than any other number. The probability of obtaining 6 heads is .2256. There is only 1 path with all heads and the resulting probability is only .0002.

Finally, we shall use our program to answer a question

raised by E. Sprinchorn relating to a bet that appears in Act 5, Scene 2 of Shakespeare's play "Hamlet." The bet is reported as follows:

> The King sir, hath laid, Sir, that in a dozen passes between yourself and him, Laertes shall not exceed you three hits. He hath laid on twelve for nine.

Sprinchorn interprets the "twelve for nine" as the King giving 12 to 9 odds in favor of Laertes winning the match. The relation between giving odds that an event will occur and assigning a probability p that the event will occur is given by the following definition.

DEFINITION. The <u>odds</u> in favor of an event E are r:s (r to s), if the probability of the event is p, and $r/s = p/(1-p)$. Any two numbers with this same ratio may be used in place of r and s.

By this definition if odds of r:s are given that an event with probability p will occur then $r(1-p) = sp$ or $p = r/(r+s)$. Thus the King giving 12 to 9 odds in favor of Laertes winning the match would correspond to his assigning a probability $p = 12/21 = .5714$ for Laertes winning or .4286 for Hamlet winning.

The difficulty comes in interpreting the condition "shall not exceed you three hits." Sprinchorn makes a case for the following two interpretations:

(a) To win, Laertes must score three more hits than Hamlet

in twelve passes.

(b) To win, Laertes must score three successive hits before Hamlet scores three such hits with the match ending after 12 passes.

In each case it is assumed that on each pass one but not both achieve a hit.

Whenever there are two ways to interpret a problem in probability we can find the answer under each interpretation. To do this we modify the program BINARY to run through all possible ways the 12 passes may go and count the sequences in which Laertes would win. The proportion of the total will then give the probability that he wins.

The program SHAKESP computes the probability under interpretation (a), i.e., that Laertes has simply to win at least three more hits than Hamlet. Recall that in the program BINARY1 we represented the outcome of a toss by the remainder R which was a 0 or a 1. We now change this to a -1 or +1 by changing R to 2*R-1 in line 180. Then Z, computed in this line, keeps track of the score by counting a 1 for a hit and a -1 for a miss. In line 210 we check if the score was $\geq$ 3, i.e., Laertes wins. W in line 230 keeps track of the number of sequences in which Laertes wins.

```
SHAKESP

100 READ N
110 DATA 12
120 FOR A = 0 TO 2↑N-1
130    LET Z = 0
140    LET M = A
150    FOR J = 1 TO N
160       LET D = INT(M/2)
170       LET R = M-2*D
180       LET Z = Z + 2*R-1
190       LET M = D
200    NEXT J
210    IF Z >= 3 THEN 230
220    GO TO 240
230    LET W = W+1
240 NEXT A
250 PRINT W,2↑N,W/2↑N
260 END
```

```
SHAKESP

794            4096            .193848
```

We note from the run of the program that Laertes would win in 794 of the 4096 possible sequences for the passes. If we assume players are of equal skill this would give a probability of .194 for Laertes to win which would mean that the odds of 12 to 9 would not be at all appropriate.

We next modify the program SHAKESP to make the same computation under the second interpretation. The program SHAKES1 is such a program. We return to representing the outcomes as 0 or 1 and we look for runs of three 0's or three 1's. Such a run would show up by a sum of 0 or 3 for three successive outcomes. We check for this in lines 220 and and 230. If a sum of three occurs we record in line 260 a win for Laertes and go to the next game. If a sum of 0 occurs Hamlet wins and we simply go to the next game.

```
SHAKES1

100 DIM H(12)
110 READ N
120 DATA 12
130 FOR A = 0 TO 2↑N-1
140    LET M = A
150    FOR J = 1 TO N
160       LET D = INT(M/2)
170       LET R = M-2*D
180       LET H(J) = R
190       LET M = D
200    NEXT J
210    FOR I = 1 TO N-2
220       IF H(I) + H(I+1) + H(I+2) = 3 THEN 260
230       IF H(I) + H(I+1) + H(I+2) = 0 THEN 270
240    NEXT I
250    GO TO 270
260    LET W = W+1
270 NEXT A
280 PRINT W,2↑N,W/2↑N
290 END

SHAKES1

1815            4096            .443115
```

We see now that the probability that Laertes wins is .443 which is quite consistent with the odds of 12 to 9 which lead to a probability of .4286.

The author was introduced to the duel problem in Hamlet by a paper written by a Dartmouth student Daniel M. Cobb. Mr. Cobb independently arrived at this problem and came up with an equally interesting and quite different interpretation of the duel. Under his interpretation the King gives odds of 12 to 9 in favor of Hamlet winning. To win Hamlet must score three consectutive hits in twelve bouts and Laertes wins if he prevents this. In Exercise 16 you are asked to find the probability of Hamlet winning to compare this probability with that corresponding to 12 to 9 odds in favor of Hamlet.

EXERCISES

1. Write a program to simulate ten rolls of a die and print out the results.
2. Modify the program of Exercise 1 to simulate 1000 rolls of the die. Do not print the result of each roll. At the end of the experiment print the proportion of sixes which turned up.
3. Simulate 1000 tosses of a coin which is biased so that heads turn up with probability .7 on each toss. At the end of the experiment, print the proportion of heads which turned up.
4. A point is picked at random from the unit square with corners (0,0),(0,1),(1,0),(1,1). Define the probability that the outcome lies in a region of area A to be A.

   (a) What is the probability that the outcome has an x coordinate less than 1/2 ?

   Ans. 1/2

   (b) What is the probability that the outcome lies in a circle with center (1/2,1/2) and radius 1/2 ?

   Ans. $\pi/4$
5. (a) Write a program to pick a point at random from the unit square defined in Exercise (4).

(b) Using (a) choose 10000 points at random, keeping track of the proportion p which lie in the circle defined in Exercise 4. What theoretical value does p approximate?

6. Show that there are 6 different triples of numbers between one and six which yield a sum of 9 and the same number which yield a sum of 10. Explain why, despite this fact, the probability of a sum of 9 is different then a sum of 10 when three die are thrown.

7. A roulette wheel is marked with green slots 0 and 00, even numbered red slots from 2 to 36, and odd numbered black slots from 1 to 35.

   (a) Describe the sample space for an experiment consisting of one spin of the wheel.

   (b) Write a computer program to simulate 1000 spins of the roulette wheel and keep track of the proportion of times a red number is spun.

   (c) In roulette if you bet on red, you win one dollar if a red number turns up, lose one dollar if a black number turns up. If a 0 or a 00 turns up play continues until either a red or a black number turns up. If it is red you break even. If it is black you lose one dollar. Modify your program in (b) to keep track of your fortune if you bet each time on red.

8. Modify the program GALILEO to simulate 4 rolls of a die and record if at least one 6 occurs. Use this to estimate the

probability of at least one six occuring when four dice are rolled.

9. In matching pennies a player starts with 0 and his fortune increases by 1 each time a head turns up and decreases by 1 each time a tail turns up. Modify the program BINARY so that it prints out a player's fortune in three penny matches for all possible ways that the game may go.

10. Modify your program for Exercise 9 to find the probability that in twelve penny matches a player's fortune never becomes negative. Compare this probability with the probability that he wins exactly 6 times.

11.* Modify the program BINARY to compute the N digit representation to the base R of the numbers between 0 and R^N-1 (BINARY does this for R = 2) Show that this can be used to compute probabilities relating to a sequence of N rolls of a die by taking R = 5. Modify the program BINARY1 to compute the probability that the sum of N rolls of a die is J. Run the program for N = 3 and check the results given in the introduction for the probabilitiy of obtaining a 9 or a 10 for the sum of three rolls. Show that this can be used to compute probabilities relating to a sequence of N rolls of a die by taking R = 5.

12. A man is willing to give 5:4 odds that his candidate will win a certain political race. What must be the probability that his candidate wins for this to be a fair bet?

Ans. 5/9.

13. A woman offers 1:3 odds that event E will occur, 1:2 odds that event F will occur. She knows that events E and F cannot both occur. What odds should she give that E or F will occur?

Ans. 7:5.

14. What odds should a person give on a bet that at least one head will turn up when two coins are tossed?

Ans. 3:1.

15. A woman offers to bet "dollars to doughnuts" that it will rain tomorrow. Assuming that a doughnut costs five cents, what must the probability be for rain to make this a fair bet?

Ans. 20/21.

16. Recall that Mr. Cobb interpreted the duel problem from Hamlet as follows: The King is giving odds of 12 to 9 in favor of Hamlet winning. To win Hamlet must score three consecutive hits in twelve bouts and Laertes will win if he prevents Hamlet from doing this. Modify the program SHAKESP to find the probability that Hamlet will win under this interpretation if the two are of equal skill. How does this probability compare with the probability corresponding to 12 to 9 odds in favor of Hamlet winning?

4. A COUNTING TECHNIQUE

We have seen that one important class of probability measures arises in experiments where all possible outcomes are assigned the same probability. In this case the problem of finding the probability of any event relating to the experiment reduces to a counting problem. In our study of probability theory we shall see that the equiprobable measure is appropriate only in a very small class of experiments. However, the counting techniques developed in analyzing these experiments will be basic to our general study of probability theory.

Many counting problems arise from counting the number of possible outcomes of an experiment which take place in several stages.

Example 1. A man wants to travel from Hanover, New Hampshire, to San Francisco, California. He must first go from Hanover to Boston. He can do this in three ways: by car, bus, or plane. From Boston he can fly or take the train to Chicago. From Chicago he must fly to San Francisco and has a choice of three flights, United American or TWA. We wish to know the number of different ways he can make the trip. In Figure 7 we have drawn a tree to show the possiblities.

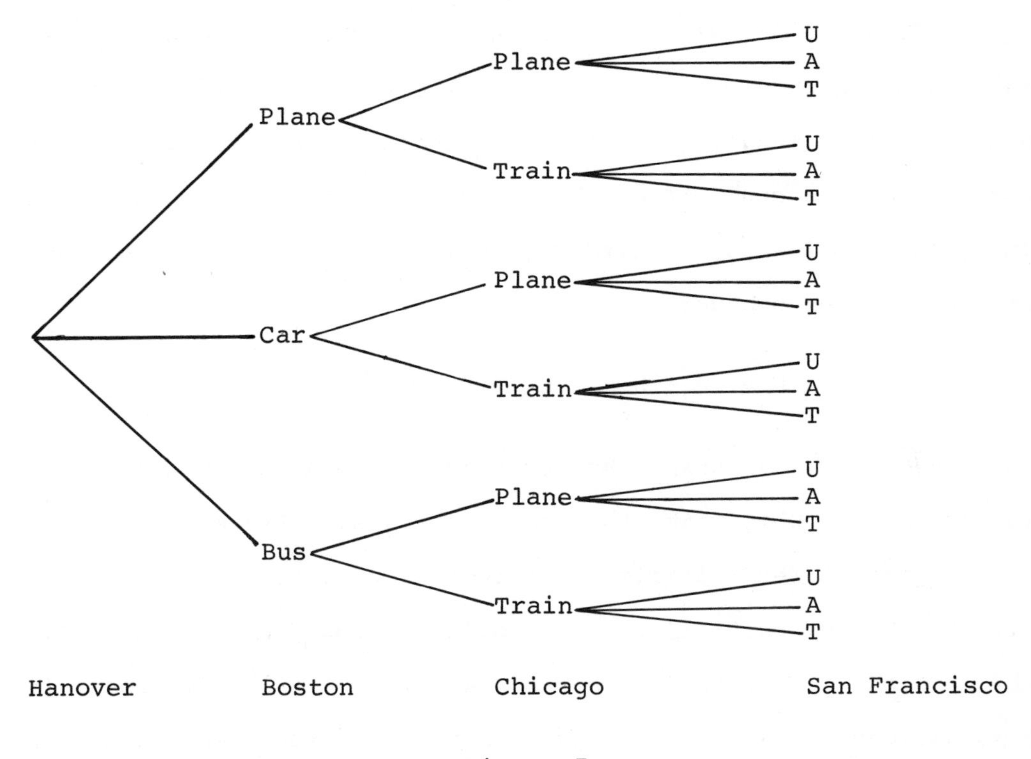

Figure 7.

There are three ways to accomplish the first stage of his trip, two ways to accomplish the second, and three ways to accomplish the third. The number of ways he can travel from Hanover to San Fransisco is given by the product of these ways: 3x2x3 = 18. In this example we can also see by counting the paths through the tree that there are indeed 18 ways to make the trip. This example is a special case of a general counting principle.

A COUNTING TECHNIQUE. Assume that a task is to be carried out in a sequence of r stages. There are n_1 ways to carry out the first stage. For each of these there are n_2 ways to carry out the second stage, then n_3 ways for the third, etc. The total number of ways in which the entire task can be accomplished is given by the product

$$N = n_1 n_2 n_3 \cdots n_r.$$

Example 2. We shall show that there are at least two people in Columbus, Ohio with the same initials. Assume that each person has three initials. There are 26 possibilities for a person's first initial, then 26 for the second, and again 26 for the third. Therefore there are $26^3 = 17{,}576$ possible combinations of initials. This number is smaller than the number of people living in Columbus and hence there must be at least two people with the same initials.

We next consider an example which is often used to show that one's intuition cannot always be trusted.

Example 3. (THE BIRTHDAY PROBLEM) The famous birthday problem is to find the smallest number of people one should ask to make it an even chance that at least two people have the same birthday. To solve this problem we shall first find the probability that in a group of r people no two have the same birthday. Let q be be this probability. Then the probability that at least 2 have the same probability is $p = 1 - q$. We shall take as sample space Ω all sequences of length r with each element one of the 365 days in the year (we ignore leap years). Since there are 365 choices for the first element of the sequence, for each of these 365 for the next etc. Thus there are 365^r elements in Ω. We must find the number of sequences which have no duplications of birthdays. For such a sequence we can choose any of the 365 days for the first element and then any of 364 for the second, 363 for the third, etc. until we make r choices. For the rth choice we will have $365-r+1$ possibilities. Hence the total number of sequences with no duplications is $(365)(364)(363)\ldots(365-r+1)$. Thus assuming each sequence is equally likely,

$$q = \frac{(365)(364)(363)\ldots(365-r+1)}{365^r}.$$

This is an easy expression to compute. To do this we rewrite q in the form

$$q = \frac{(365)}{(365)}\frac{(364)}{(365)}\frac{(363)}{(365)}\cdots\cdots\frac{(365-r+1)}{365}.$$

The program BIRTHD carries out this computation and prints the probabilities P for R = 5 to 75 every 10 units.

```
BIRTHD

10 PRINT " R", "    P"
20 PRINT
30 FOR R = 5 TO 75 STEP 10
35    LET Q = 1
40    FOR J = 1 TO R
50       LET Q = Q*(365-J+1)/365
60    NEXT J
70    PRINT R,1-Q
80 NEXT R
100 END
```

```
BIRTHD

R                  P

5               .024470
15              .250854
25              .567518
35              .813875
45              .940814
55              .986225
65              .997677
75              .999719
```

We note the surprising fact that for r = 25 it is already better than an even chance that there will be at least one duplication. To see exactly where the change is we rerun the the program going from 20 to 25

```
BIRTHD

N              P

20             .409826
21             .442164
22             .474259
23             .505947
24             .537079
25             .567518
```

We see that for r = 22 the probability is less than 1/2 and for r = 23 it is greater than 1/2.

We now consider a somewhat more general problem: the problem of the number of ways that we can order a set of n objects. We are going to put n objects in a row. For the first position there are n choices, namely any of the objects. For each of these choices there are n-1 choices for the second position. Then there are n-2 choices for the third etc. This process continues until there is one place remaining and one object remaining. The last place can be filled in only one way. The number of arrangements, or permutations, of n objects is

$$n(n-1)(n-2)....1.$$

This number is called "n factorial", and is denoted by n!. For example, 3! = 3x2x1 = 6, 4! = 4x3x2x1 = 24, etc.

The numbers n! become large very fast. The program FACTOR prints out these numbers for n = 1 to 10.

```
FACTOR

100 PRINT " N"," N FACTORIAL"
110 PRINT
120 FOR N = 1 TO 10
130    LET X = 1
140    FOR I = 1 TO N
150       LET X = I*X
160    NEXT I
170    PRINT N,X
180 NEXT N
190 END
```

```
FACTOR

N               N FACTORIAL

1               1
2               2
3               6
4               24
5               120
6               720
7               5040
8               40320
9               362880
10              3628800
```

The number 10! = 3628800 is sometimes taken to be the number of operations which it is reasonable to ask a computer to do. It might be remarked that Leibnitz in 1666 computed

$$24! = 6,204,484,017,332,394,339,360,000$$

though presumeably in less than 10! operations.

While we shall not make heavy use of it there is an important approximate formula for the numbers n! called <u>Stirling's formula</u>. This is

$$n! \doteq \sqrt{2\pi n}\, n^n e^{-n}.$$

We shall use the symbol $\doteq$ to mean "approximately equal to."

Stirling's formula is approximate in the sense that the ratio of the true value to its approximation tends to 1 as n tends to infinity. In Exercise 9 you are asked to write a program to illustrate this approximation.

There is a second famous nonintuitive problem in probability which relates to permutations.

Example 4. (THE HAT CHECK PROBLEM) n people check their hats and because of a large wind the hats become completely mixed up before they are handed back. The problem posed is to find the probability that no one gets their own hat back. We first compute the probability in the special case of three people. Let us number the hats 1,2,3 and assume that the order in which they should be handed back is 1,2,3. The possible orders in which they can be handed back are

123
132
312
321
231
213

Only in two permutations 312 and 231 is it true that no person gets their own hat. By "at random" we mean that all ways of handing the hats back are equally likely. Then the probability that no person gets their own hat back is P(312,231) = 2/6 = 1/3.

In general we shall have to count the number of permutations which do not leave any element fixed. Such a permutation is called a complete permutation. For the case of three elements we have seen that 312 and 231 are the only complete permutations. For example 213 is not complete since 3 remains fixed under this permutation.

Turning now to the general case of n hats we see that there are $n!$ permutations. Let $w(n)$ be the number of complete permutations. The probability that no one gets their own hat back is $w(n)/n!$. Thus we need to find $w(n)$.

We shall do this by developing a recursion relation to generate the values of $w(n)$ from initial values. We describe the relation first in a special case. Assume that we want to find the number of ways that 10 people can have hats none of which are their own. Consider one person, say Ann Arbor. Ann arrives with her own hat. If every other person has a hat different than their own, she can choose any of the nine and change hats with that person and noone will have their own hat. By assumption their are $w(9)$ ways that the nine can have diffent hats. She has 9 choices for person with whom she changes hats. Thus we have $9w(9)$ ways that this can be done. However, their are other possibilities. Assume that she comes upon the nine people and one has his own hat and all the other eight have different hats. Then she has only to choose the person with their own hat and change with that person and again all ten will have different hats. This could occur in $9w(8)$ different ways. There is no duplication in these two ways that Ann can give up her hat, since if we were to undo the process, in one case we would end up with 1 person having his own own hat and in the other case 2. Also it is clear that these are the only two ways that Ann can give up her hat so that all have different hats. Thus

$$w(10) = 9w(9) + 9w(8).$$

But the above argument applies equally well for the case of n persons and gives the relation

$$w(n) = (n-1)w(n-1) + (n-1)w(n-2).$$

Dividing through by $n!$ we have

$$\frac{w(n)}{n!} = \frac{(n-1)}{n!} w(n-1) + \frac{(n-1)}{n!} w(n-2)$$

$$= \frac{(n-1)}{n} \frac{w(n-1)}{(n-1)!} + \frac{1}{n} \frac{w(n-2)}{(n-2)!}$$

Thus we have

$$(1) \quad p_n = \frac{(n-1)}{n} p_{n-1} + \frac{1}{n} p_{n-2}.$$

We can compute successive values of p_n from (1) if we have two initial values to start with. If we have 1 person this person must have his own hat and hence $p_1 = 0$. If we have two people there are two possibilities, either they both have their own hats or they have each others. Thus $p_2 = 1/2$. Hence we choose these two values as initial values. The program HAT then uses (1) to generate the values from 3 to 10.

```
HAT

100 PRINT " N", "PROBABITY THAT NO"
110 PRINT "                      PERSON GETS OWN HAT"
120 PRINT
125 LET P(1) = 0
126 LET P(2) = 1/2
140 FOR N = 3 TO 10
150     LET P(N) = ((N-1)/N)*P(N-1) + (1/N)*P(N-2)
210     PRINT N,P(N)
220 NEXT N
230 END
```

```
HAT

N              PROBABITY THAT NO
               PERSON GETS OWN HAT

3               0.333333
4               0.375
5               0.366667
6               0.368056
7               0.367857
8               0.367882
9               0.367879
10              0.367879
```

We note that the probabilities p_n alternately increase and then decrease. Further these numbers seem to be tending to a limiting value and in fact it would seem that the probability is going to be essentially the same after we have more than seven people. We can see that this is indeed the case by the following further analysis of the values of p_n. From (1) we see that

$$(p_n - p_{n-1}) = \frac{1}{n}(p_{n-2} - p_{n-1}).$$

If we let

$$v_n = p_n - p_{n-1}$$

we have

$$(2)\quad v_n = -\frac{1}{n} v_{n-1}.$$

We know that $v_2 = p_2 - p_1 = 1/2$. From (1) we obtain

$$v_3 = -\frac{1}{3x2}$$

$$v_4 = \frac{1}{4x3x2}$$

and in general

$$v_n = (-1)^n \frac{1}{n!}.$$

Next we write

$$p_n = (p_n - p_{n-1}) + (p_{n-1} - p_{n-2}) + \dots + (p_2 - p_1) + p_1$$

Since $p_1 = 0$,

$$p_n = v_1 + v_2 + \dots + v_n$$

$$= \frac{1}{2!} - \frac{1}{3!} + \frac{1}{4!} - \dots \pm \frac{1}{n!}.$$

Those that have had calculus will recall that

$$e^x = 1 + x + \frac{x}{2!} + \frac{x}{3!} + \dots$$

so that

$$e^{-1} = \frac{1}{2!} - \frac{1}{3!} + \frac{1}{4!} + \ldots$$

Thus p_n is the nth term of the series for e^{-1} and tends to e^{-1} = .3678794... as n tends to infinity.

EXERCISES

1. Four people are to be arranged in a row to have their picture taken. In how many ways can this be done?

2. An automobile manufacturer has four colors available for automobile exteriors and three choices for interiors. How many cars of different color combinations can he produce?

3. Assume that a typical computation on the computer takes a microsecond (one millionth of a second). Estimate the time required to carry out a program which makes 10! computations.

4. List all possible arrangements of the first 3 numbers. Describe a systematic way to obtain all arrangements of the first 4 numbers from this list.

5. There are three different routes connecting city A to city B. How many ways can a round trip be made from A to B and back? How many ways if it is desired to take a different route on the way back?

   Ans. 9,6

6. In arranging people around a circular table we take into

account their seats relative to each other, not the actual position of any one person. Show that n people can be arranged around a circular table in $(n-1)!$ ways.

7. Five people get on an elevator which stops at five floors. Assuming that each has an equal probability of going to any one floor find the probability that they all get off at different floors.

Ans. .0384

8. A finite set Ω has n elements. Show that if we count the empty set and Ω as subsets there are 2^n subsets of Ω.

9. Let A_n be the approximation for $n!$ as given by Stirling's formula. Write a program which prints out $n!$, A_n, $A_n/n!$, and $A_n - n!$ for $n = 1$ to 9.

10. A more refined inequality for approximating is given by

$$\sqrt{2\pi n}\, n^n e^{1/(12n+1)} < n! < \sqrt{2\pi n}\, n^n e^{-n} e^{1/(12n)} .$$

Write a computer program to illustrate this inequality for $n = 1$ to 9.

11.* Write a computer program to generate all permutations of n objects for $n = 1$ to 6.

12. Modify the program of Exercise 11 to compute the probability that in the hat check problem exactly k out of n people get their own hat back for $n = 1$ to 6 and $k = 1$ to n.

13. For the hat check problem, derive a relation between the

probability that a person gets his own hat with n people and the probability that exactly 1 person get his own hat in the case of n+1 people.

14. In a well-known game of solitaire a player turns over each card in a deck of cards one-by-one while giving their names in a fixed order. The objective is to go through the deck without once naming the card turned over. What is the probability of winning? What is the probability if the person uses only the 13 spades?

15. Find a formula for the probability that among a set of N people at least two have their birthdays in the same month of the year.

16. Consider the problem of finding the probability of more than one coincidence of birthdays in a group of n people. For example, three people with the same birthday or two pairs of people with the same birthday or larger coicidences. Show how you could compute this probability and write a computer program to carry out this computation. Use your program to find the smallest number of people for which it would be a fiar bet that there would be more than one coincidence of birthays.

5. BINOMIAL COEFFICIENTS

Let U be a set with n elements. We wish to compute the number of subsets of U that have exactly j elements. We shall consider the empty set ∅ and the entire set U to be subsets of U.

To form a subset of U, consider each of the n elements of the set. For each element, we have two choices: either the element is to be included in the subset or it is not. We can regard the problem of determining the number of subsets as an n stage process with 2 choices at each stage. Thus, by the results of the last section, the total number of subsets of U is 2^n.

Example 1. Let U = {A,B,C}. There are $2^3 = 8$ subsets of U:

∅

{A}

{B}

{C}

{A,B}

{A,C}

{B,C}

{A,B,C}

Note that there is only one subset with no elements, 3 subsets with exactly one element, 3 with exactly 2 elements, and 1 subset with exactly 3 elements.

We consider now the more general problem of finding the

number of subsets with j elements that can be formed from a set U of n elements. We call this number a binomial coefficient and denote it by C(n,j). A more common notation for C(n,j) is $\binom{n}{j}$. We shall use both notations. The matrix notation C(n,j) is more convenient for use in programs which use these numbers. The terminology binomial coefficient comes from an application to the binomial theorem which we shall discuss later.

We note first that C(n,n) = 1. That is, there is only one subset of U which contains all n elements of U. Further, C(n,0) = 1. That is, there is only one subset of U having no elements. The empty set has one subset, namely the empty set. Therefore C(0,0) = 1. The remaining values of C(n,j) are determined by the following relation.

THEOREM. For $0 < j < n$ the binomial coefficients satisfy:

$$(1)\quad C(n,j) = C(n-1,j-1) + C(n-1,j).$$

Proof. To see that (1) is true, let U be a set with n elements. We choose one element r from of the set U. In choosing a subset of j elements, we have two choices for the element r: either we want it in our subset or we do not. If we want r in the subset, we must choose j-1 elements out of the remaining n-1 elements to complete our subset. This can be done in C(n-1,j-1) ways. If we do not want r in the subset we must choose j elements out of the remaining n-1 elements. This can be done in C(n-1,j) ways. Since r is either in the chosen subset or is not, the sum of the two numbers gives the number of subsets of j elements that can be chosen from a set of n elements, i.e., $C(n,j)$. That is, the relation (1) is satisfied.

The relation (1) together with the knowledge that

$$C(n,0) = C(n,n) = 1$$

determine completely the numbers $C(n,j)$. To see this, we construct the famous Pascal triangle shown in Figure 8.

1									
1	1								
1	2	1							
1	3	3	1						
1	4	6	4	1					
1	5	10	10	5	1				
1	6	15	20	15	6	1			
1	7	21	35	35	21	7	1		
1	8	28	56	70	56	28	8	1	
1	9	36	84	126	126	84	36	9	1

Figure 8.

The nth row of this triangle has the entries $C(n,0), C(n,1), \ldots, C(n,n)$. We know that the first and last of these numbers are 1. This determines the first two rows. The remaining rows are determined by (1). That is, the entry $C(n,j)$ for $0 < j < n$ is the sum of the entries immediately above and above and one to the left. For example $C(5,2) = 10 = 6 + 4$.

The algorithm by which the table is constructed makes it an

easy task to write a computer program to compute the binomial coefficients. The program PASCAL is such a program.

```
PASCAL

10 DIM C(50,50)
20 FOR N = 0 TO 50
30    LET C(N,0) = C(N,N) = 1
40    FOR J = 1 TO N-1
50       LET C(N,J) = C(N-1,J-1) + C(N-1,J)
60    NEXT J
70 NEXT N
```

We have not included an END statement or any PRINT instructions. The program will not run as it stands. We use it instead as a part of any other program that requires the binomial coefficients.

While the Pascal triangle gives a way to construct the binomial coefficients it is also possible to give a formula for $C(n,j)$, namely

$$(2)\quad C(n,j) = \frac{n!}{j!(n-j)!}.$$

To see that formula (2) is correct we use the following argument. We know that there are $n!$ possible arrangements of the n objects, and we can obtain these in the following manner. Choose j objects and order them. This may be done in $j!$ ways. Then we can order the remaining $n-j$ objects in $(n-j)!$ ways. Since there are $C(n,j)$ ways to choose j objects out of n, the total number of arrangements of $n!$ is determined by

$$n! = C(n,j)j!(n-j)!.$$

Thus

$$C(n,j) = \frac{n!}{j!(n-j)!}.$$

We shall be primarily interested in the application of binomial coefficients to probability but we shall need to remind the reader of one application to algebra. This is the binomial expansion

$$(a+b)^n = \sum_{j=0}^{n} C(n,j)a^j b^{(n-j)}.$$

For example,

$$(a + b)^2 = a^2 + 2ab + b^2$$
$$(a + b)^3 = a^3 + 3a^2b + 3ab^2 + b^3$$

etc.

The term binomial coefficients for $C(n,j)$ was suggested by this applications.

THEOREM. The probability of obtaining exactly j heads in n tosses of a fair coin is given by

$$B(n,j) = C(n,j)/2^n.$$

Proof. We recall that in n tosses of a fair coin we represented the sample space as the space of all sequences of the form

$$\omega = (H,H,T,\ldots,H,T)$$

To find the probability of the event "heads on exactly j out of n tosses" we count the number of such sequences which have exactly j H's and $(n-j)$ T's. But this is just the number of ways that we can choose j of the positions to put a H and $n-j$ to put a T. That is, $C(n,j)$. There are 2^n possible sequences of length n as so we assign an equal probability of $1/2^n$ to each such sequence. Thus the theorem is proven.

We next show that the number $B(n,j)$ satisfy a relation similar to (1). We note first that there is only one sequence for which all outcomes are H and similarly there is only one for which no outcomes are H. Thus

$$B(n,n) = B(n,0) = 1/2^n.$$

For $0 < j < n$ we have from (1),

$$C(n,j)/2^n = C(n-1,j-1)/2^n + C(n-1,j)/2^n$$

But since $B(n,j) = C(n,j)/2^n$ this is equivalent to

$$B(n,j) = B(n-1,j-1)/2 + B(n-1,j)/2.$$

Using this relation we can by making only minor changes in the program PASCAL write the program BINOM to compute B(n,j).

```
BINOM

5 DIM B(50,50)
10 DEF FNF(X) = INT(1000*X+.5)/1000
20 FOR N = 0 TO 50
30    LET B(N,0) = B(N,N) = 1/2↑N
40    FOR J = 1 TO N-1
50       LET B(N,J) = 1/2*B(N-1,J-1) + 1/2*B(N-1,J)
60    NEXT J
70 NEXT N
80 FOR N = 0 TO 9
90    FOR J = 0 TO N
100       PRINT FNF(B(N,J));
110    NEXT J
120    PRINT
130 NEXT N
200 END
```

```
BINOM

 1
.5    .5
.25   .5    .25
.125  .375  .375  .125
.063  .250  .375  .250  .063
.031  .156  .313  .313  .156  .031
.016  .094  .234  .313  .234  .094  .016
.008  .055  .164  .273  .273  .164  .055  .008
.004  .031  .109  .219  .273  .219  .109  .031  .004
.002  .018  .07   .164  .246  .246  .164  .07   .018  .002
```

We have given a listing of this program and a run which computes the values of B(n,j). The rows give the values for successive values of n. We note that going across a row,i.e., for fixed n, that the values are small near the extreme values and largest in the middle. We are now in a position to compute these values for larger n and graph the probabilities for fixed n. The program BINOM1 carries this out for n = 20. We see that the graph is in fact a bell shaped curve. The explanation of this curve will come later when we study the celebrated central limit theorem of probability.

```
BINOM1

10 DIM B(50,50)
20 FOR N = 0 TO 50
30    LET B(N,0) = B(N,N) = 1/2↑N
40    FOR J = 1 TO N-1
50       LET B(N,J) = .5*B(N-1,J-1) + .5*B(N-1,J)
60    NEXT J
70 NEXT N
80 FOR J = 0 TO 20
90    PRINT TAB(100*B(20,J));"*"
100 NEXT J
200 END
```

```
BINOM1

*
*
*
*
*
*
  *
      *
           *
               *
                *
               *
           *
      *
  *
*
*
*
*
*
*
```

Note that in line 90 we have used the TAB function. When we insert TAB(M) in a PRINT statement the teletype "tabs" to column

M, and prints the value there. On the normal teletype there about 75 possible columns so that we have to adjust M so that the values encoutered lie between 0 and 75. It is for that reason that we have multiplied B(20,J) by 100 in line 90.

EXERCISES

1. Compute the following:

(a) C(6,3)

(b) B(5,4)

(c) C(7,2)

(d) C(26,26)

(e) B(8,4)

(f) $\binom{6}{2}$

(g) $\binom{10}{9}$

2. In how many ways can we choose five people from a group of ten to form a committee?

3. A fair coin is thrown ten times. What is the probability that seven or more heads turn up?

4. How many seven-element subsets are there in a set of nine elements?

5. If a set has 2n elements, show that it has more subsets with n elements than with any other number of elements. (Hint: Consider the ratio C(2n,i)/C(2n,i-1) for i = 1,2,...,n.)

6. Let B(2n,n) be the probability that in 2n tosses of a fair coin exactly n heads turn up. Using Stirling's formula show that B(2n,n) $\doteqdot \frac{1}{\sqrt{\pi n}}$. Using the program BINOM compare

this with the exact value for $n = 10$ to 25.

7. A poker hand is a set of five cards randomly chosen from a deck of 52. Find the probability of

   (a) royal flush (ten,jack,queen,king, ace in a single suit).

   Ans. $4/(C(52,5) = .0000015$.

   (b) straight flush (five in a sequence in a single suit, and not a royal flush).

   Ans. $(40-4)/C(52,5) = .000014$.

   (c) four of a kind (four cards of the same face value).

   Ans. $624/C(52,5) = .00024$.

   (d) full house (one pair and one triple of the same face value).

   Ans. $3744/C(52,5) = .0014$.

   (e) flush (five cards in a single suit and not a straight or royal flush).

   Ans. $(5148-40)/C(52,5) = .0020$.

   (f) straight (five cards in a row, not all the same suit).

   Ans. $(10240-40)/C(52,5)=.0039$.

8. Show that $C(n,j) = C(n,n-j)$.

9. A lady wishes to color her fingernails on one hand using at most 2 of the colors red,yellow and blue. How many ways can she do this? (Hint: Your answer is three too large.)

10. A football team plays 8 games in a particular season,

winning 3, losing 3 and ending 2 in a tie. Show that the number of ways that this can happen is $C(8,3)C(5,3) = 8!/(3!3!2!)$.

11. Using the technique of Exercise 10 show that the number of ways that one can put n objects into three boxes with a in the first b in the second and c in the third is $n!/(a!b!c!)$

12. Proof the following "Binomial identity"

$$C(2n,n) = \sum_{j=0}^{n} C(n,j)C(n,n-j).$$

13. A fair (6-sided) die is thrown five times. What is the probability that: (a) All throws are different? (b) The sum of all throws is no greater than 6? (c) Exactly three sixes and exactly two fives are thrown? (d) Exactly three sixes are thrown?

15. A gin hand consists of 10 cards from a deck of 52 cards. Find the probability that a gin hand: (a) has all 10 cards of the same suit. (b) has 4 cards in one suit, 3 in another, 2 in another, and 1 in a fourth suit; (c) includes all 4 cards of each of the two different face values; (d) includes 3 cards in each of two different face values and less than 3 cards of any other face value.

14*. Using the program PASCAL compute the first 7 rows of the Pascal triangle printing out the values of the binomial coefficients mod 2. The sum of the rows will give the number of odd numbers in the row. Verify that for the nth row this is $2^{b(n)}$ where $b(n)$ is the number of 1's in the binary expansion of n. Can you prove this? You might try to prove first that the rows of the form $2^n - 1$ have only odd terms.

6. CONDITIONAL PROBABILITY

Assume that we have assigned a probability measure P to a sample space $\Omega = \omega_1, \omega_2, \ldots, \omega_n$. We now would like to compute the probability that an event A occurs given that the event E occurs. We assume $P(E) > 0$. We denote by $P(A|E)$ the probability of the event A given that E occurs. How should we assign these probabilities? Clearly if ω_j is a sample point not in E we want $P(\omega_j|E) = 0$. Secondly, we want the probabilities for ω_k in E to have the same relative probabilities that they had before we learned that E happens. That is, for some $c > 0$

$$P(\omega_k|E) = cP(\omega_k)$$

for all ω_k in E. But we must also have

$$(1) \qquad \sum_E P(\omega_k|E) = c \sum_E P(\omega_k) = 1.$$

Thus

$$(2) \quad c = \frac{1}{\sum_E P(\omega_k)} = \frac{1}{P(E)}.$$

This means we should define

$$P(\omega_k|E) = \frac{P(\omega_k)}{P(E)}$$

for ω_k in E. For a general event A we then have

$$(3) \quad P(A|E) = \sum_{AE} P(\omega_k|E) = \frac{P(AE)}{P(E)}.$$

We can also rewrite this as

$$P(AE) = P(A)P(E|A).$$

Example 1. Let us return to the example of three candidates A, B, C running for office. We assigned $P(A) = P(B) = 2/5$ and $P(C) = 1/5$. Assume now that A drops out of the race. Let E be the event "B or C wins". Then $P(A|E) = 0$,

$$(4)\quad P(B|E) = \frac{2/5}{2/5 + 1/5} = 2/3$$

and $P(C|E) = 1/3$. We note that B is still twice as likely as C is to win.

When we analyze an experiment which takes place in a sequence of steps we shall often find it convenient to represent the sample space as the set of all paths through a tree. We assign a probability measure by first considering the conditional probabilities that we wish for the outcome of the jth step, given all previous outcomes. We assign these as weights at the appropriate branchs of the tree. We then choose, as weights for a path through the tree, the product of these branch weights along the path. This yields a probability measure with the desired conditional probabilities. We call this a _tree measure_. We illustrate this procedure by several examples.

Example 2. We have two urns, I and II. Urn I contains 2 black balls and 3 white balls. Urn II contains 1 black and 1 white ball. An urn is chosen at random and a ball is chosen at random from it. We can represent the sample sapce as the paths through a tree as shown in Figure 9.

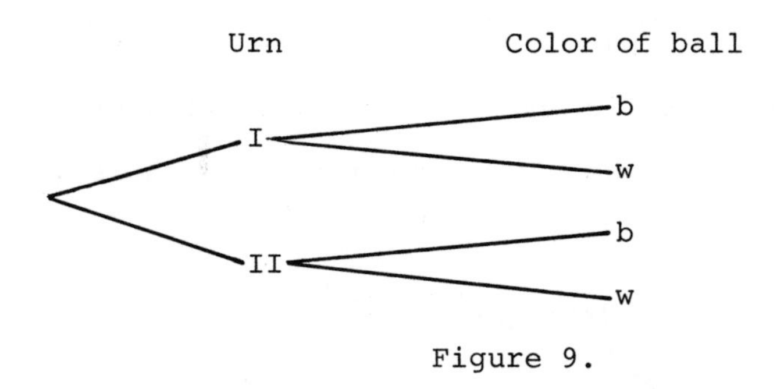

Figure 9.

In this example our experiment takes place in two stages. We have information which suggests the appropriate probabilities for the outcome of the first stage and then the conditional probabilities for the outcomes of the second stage given the outcome of the first stage. We indicate these on the branches of the tree in Figure 10.

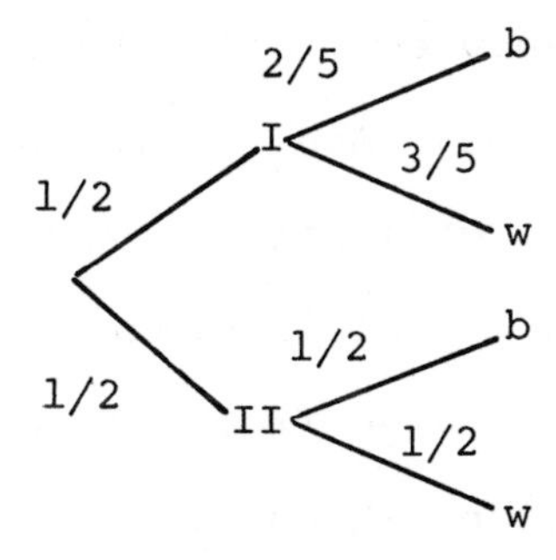

Figure 10.

Thus we assign

$$P(I,b) = P(I)P(b|I) = (1/2)(2/5) = 1/5$$

$$P(I,w) = P(I)P(w|I) = (1/2)(3/5) = 3/10.$$

Thus a point of Ω is a path through the tree and we assign a probability for a specific path to be the product of the probabilities along the paths. Then our <u>tree measure</u> is given by Figure 11.

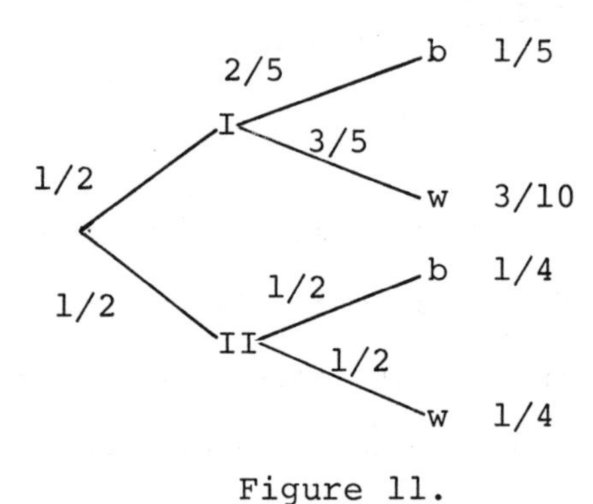

Figure 11.

<u>Example 3</u>. Three men, Abel, Baker and Charlie, are in jail and one of them is to be executed. The guard knows which is the unlucky man but Abel does not. Abel says to the guard, "I know that either Baker or Charlie will not be executed. Therefore you won't really be giving me any real information if you tell me the name of one of those who will go free. If both are to go free just toss a coin to decide which to tell me." The guard considers the request and then says, "No, that would not be fair to you. At the moment you think that you have probability $\frac{1}{3}$ of being executed. If I tell you the name of the one who will go free you will then have probability $\frac{1}{2}$ and will not sleep so well tonight." Was the guard right? To answer this, Abel examines the possible outcomes as a two stage process, the first stage being

the choice of the man to be executed and the second the guard's answer.

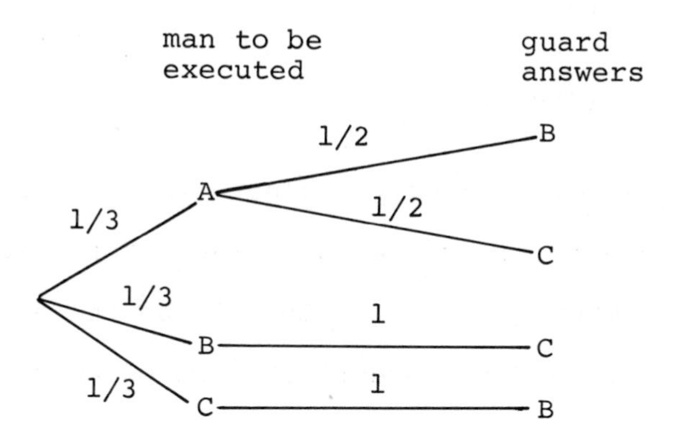

Figure 12.

Let E be the event "the guard says Baker" and F the event "Abel is to be executed". Then Able computes

$$(5)\quad P(F|E) = \frac{P(EF)}{P(E)} = \frac{1/6}{1/3 + 1/6} = 1/3.$$

By symmetry the probability that Abel is to be executed given the guard says Charlie is also 1/3. Hence the guard is wrong. His answer makes no difference to Abel.

DEFINITION. Two events E and F are said to be <u>independent</u> if

$$P(F|E) = P(F)$$

Thus we see that E and F are independent if the knowledge that E is true does not affect our prediction about F. If E and

F are independent then $P(F) = P(F|E) = P(EF)/P(E)$,

or equivalently,

$$P(EF) = P(E)P(F)$$

While it is usually intuitively clear when events are independent, this may not always be the case. For example, a coin is tossed twice. Then we choose Let F be the event "The first toss is a head" and E the event "the two outcomes are the same". Then

$$P(F|E) = \frac{P(HH)}{P(HH,TT)} = \frac{1/4}{1/2} = 1/2 = P(F),$$

That is, events E and F are independent.

Example 4. A coin is tossed twice. $\Omega = \{HH,HT,TH,TT\}$. Let $E = \{HH,HT\}$. Let F be the event "two heads turn up". That is, $F = \{HH\}$. $P(E) = 1/2$ and $P(F) = 1/4$. Then $P(F|E)$ is the probability that two heads occur given that the first toss resulted in a head. $P(F|E) = P(EF)/P(E) = (1/4)/(1/2) = 1/2$. $P(F|E)$ is not equal to $P(F)$. The events E and F are not independent.

We conclude this section with an example of computing a Bayes probability. This is simply a particular type of computation using conditional probability.

Example 5. Smith claims that he can tell the difference between beer and ale and Jones denies that he can. Smith claims more precisely that, in a series of tests to distinquish between the two, he will be right 60 percent of the time. Jones claims that Smith is just guessing i.e., that Smith will be right is only .5. Let H_0 be the hypothesis that Smith is correct, i.e., he has probability .6 of distinquishing. Let H_1 be the hypothesis that Jones is correct, i.e., Smith has probability .5 of being correct. We are going to carry out a test to try to determine which hypothesis is correct. Before any testing we shall assume that $P(H_0) = 1/2$ and $P(H_1) = 1/2$. This is called an _a priori_ probability. We now carry out one test. That is, Smith is given two glasses, one with beer and one with ale and he is to tell which has the beer and which has the ale. Then we wish to compute the new probability for each hypothesis given the information of the test. These are called _a posteriori_ probabilities. We proceed as follows:

$$P(H_0|C) = \frac{P(H_0C)}{P(C)}$$

$$= \frac{P(H_0)P(C|H_0)}{P(C)}$$

$$= \frac{P(H_0)P(C|H_0)}{P(H_0C) + P(H_1C)}$$

$$= \frac{P(H_0)P(C|H_0)}{P(H_0)P(C|H_0) + P(H_1)P(C|H_1)}$$

$$= \frac{(.5)(.6)}{(.5)(.6) + (.5)(.5)} = \frac{.6}{1.1} = .545.$$

Thus if Smith is correct on the first test we would change our *a priori* probability of .5 to an *a posteriori* probability of .545.

In a similar manner we find

$$P(H_0|W) = \frac{P(H_0)P(W|H_0)}{P(H_0)P(W|H_0) + P(H_1)P(W|H_1)}.$$

Thus if Smith is wrong on his first test,

$$P(H_0|W) = \frac{(.5)(.4)}{(.5)(.4)+(.5)(.5)} = \frac{4}{9} = .444.$$

We see that one test gives us some information but we would want to make more tests before making a final decision. The effect of another test may be found in exactly the same way except that we assume before the test that our estimate for $P(H_0)$ is is the probability computed on the basis of the outcome of the first experiment. It would be hoped that after a number of tests the result would assign a high probability to $P(H_0)$ if in fact H_0 is correct and a low probability if H_1 is correct. The program BAYES simulates this process for 100 tests assuming that H_0 is in fact correct. Of course in an actual test we would have to take very small glasses for 100 tests but our computer can handle this without the problems encountered by mortals. The program takes a sequence of samples and continually computes the updated probability that H_0 is true.

```
BAYES

100 DEF FNF(X) = INT(1000*X)/1000
110 READ A,P0,P1,N
120 DATA .5,.5,.6,100
130 FOR I = 1 TO N
140     IF RND > P0 THEN 170
150     LET A = A*P0/(A*P0 + (1-A)*P1)
160     GOTO 180
170     LET A = A*(1-P0)/(A*(1-P0) + (1-A)*(1-P1))
180     IF I <> INT (I/10)*10 THEN 200
190     PRINT I,FNF(A)
200 NEXT I
210 END
```

```
BAYES

10              .55
20              .4
30              .45
40              .401
60              .23
70              .451
80              .694
90              .862
100             .92
```

In the program BAYES, P0 is the probability of a correct reponse under the hypothesis H_0 and P1 is the probability under the hypothesis H_1. A is the initial probability we assign to hypothesis H_0 being correct. In the loop from 130 to 200 we carry out N test assuming in fact that H_0 is correct. The a posteriori probability for H_0 is computed after each test in line 150 or 170 depending upon whether the test was a success or a failure. Line 190 prints out the updated proabability for H_0 after every ten experiments. The data used in the program corresponds to the problem of trying to distinquish between the hypothesis H_0 that the true probability is .5 from the hypothesis H_1 that it is .6

when in fact H_0 is correct. It is assumed initially that the probability for H_0 is .5

We see that the probabilities for H_0 do seem to be tending to 1 giving us confidence that this procedure would lead us to the correct inference that H_0 is the true hypothesis. However, we note that if we had stopped after 60 tests we would have been led to suspect that H_1 is the correct hypothesis and have been in error. It is a problem of statistics to determine the number of experiments necessary to assure a reasonable deduction based upon the *a posteriori* probabilities. Of course we could also study this problem by further computer simulation.

EXERCISES

1. A coin is tossed three times. What is the probability that exactly two heads occur, given that:
   a. The first outcome was a head.
   b. The first outcome was a tail.
   c. The first two outcomes were heads.
   d. The first two outcomes were tails.
   e. The first outcome was a head and the third outcome was a head.
2. A die is rolled twice. What is the probability that the sum of the faces is greater than 7, given that:
   a. The first outcome was a 4.
   b. The first outcome was greater than 3.
   c. The first outcome was a 1.

d. The first outcome was less than 5.

3. A card is drawn at random from a deck of cards. What is the probability that:

   a. It is a heart given that it is red?

   b. It is higher than a 10 given that it is a heart?

   c. It is a jack given that it is red?

   d. It is a seven given that it is lower than a jack?

   e. It is an even-numbered card given that it is higher than a five?

4. What is the probability that a family of two children has

   a. Two boys given that it has at least one boy.

   b. Two boys given that the first child was a boy.

5. What is the probability that a bridge hand (13 cards) holds four aces given that:

   a. It holds at least one ace.

   b. It holds the ace of spades.

6. Prove Bayes' theorem: Let $\{E_1, E_2, \ldots,\}$ be a collection of exhaustive and mutually exclusive events. Then

$$P(E_k | A) = \frac{P(A|E_k)P(E_k)}{\sum_j P(A|E_j)P(E_j)} .$$

7. One coin in a collection of 65 has two heads, the rest are fair. If a coin chosen at random from the lot turns up heads 6 times in a row, what is the probability that it is

the two-headed coin?

Ans. .5.

8. Modify the program BAYES to simulate 100 experiments under the assumption that H_1 is the correct hypothesis. Run the program and see if the _a posteriori_ probabilities for H_0 would now seem to approach 0.

9. Suppose that on a certain street a car passes during second k (k = 0,1,...) with constant probability p. Suppose further that a pedestrian starts waiting at time 0 and can cross the street only if no car will arrive during the next three seconds. Write a BASIC program to simulate this process 10,000 times, calculating the proportion of times a pedestrian has to wait k seconds when p = 1/3. Derive theoretical values for k = 0,1,2,3,4.

10. You are given two urns and fifty balls. Half of the balls are white and half are black. You are asked to distribute the balls in the urns with no restriction placed on the number of either type in an urn. How should you distribute the balls in the urns to maximize the probability of obtaining a white ball if an urn is chosen at random and a ball drawn out at random. Verify that your result is the best possible.

11. If E,F,G are any three events prove that

$$P(EFG) = P(E)P(F|E)P(G|EF).$$

12. It is desired to find the probability that in a bridge deal that each player receives an ace. A student argues as follows. It does not matter where the first ace goes. The second ace must go to one of the other three players and this occurs with probability 3/4. Then the next must go to one of two an event of probability 1/2 and finally the last ace must go to the player who does not have an ace. This occurs with probability 1/4. The probability that all these events occur is the product (3/4)(1/2)(1/4) = 3/32. Is this argument correct? If not compute the probability correctly. (Ans. 3/32 = .09375. Correct answer is .105498. Reasoning is not correct.)

13. In a large city, a certain disease is present in about one out of every 1,000 persons. A program of testing is to be carried out using a detection device which gives a positive reading with probability .99 for a diseased person and with probability .05 for a healthy person. What is the probability that a person who has a positive reading actually has the disease?

    Ans. .019.

14. From the freshman class at a large Midwestern University it was found that 10 percent of the students failed math, 15 percent failed English and 2 percent failed both math and English. A student is selected at random from the freshman class. Are the events "the student failed math" and "the

student failed English" independent?

15. Let (x,y) be a random point in the unit ball $b(0,1) = \{(x,y): x^2 + y^2 \leq 1\}$ What is the probability which we must assign to $E = \{(s,y): x = 0\}$? Why is it impossible to consider $P(y \geq 0 \mid x = 0)$ according to our formulation of conditional probability? What measure would we like to assign?

16. A man who is terrified of being hijacked decides to carry a bomb with him whenever he travels in a plane. His reasoning is that there has never been a hijacking in which two people have been carrying bombs. Is his reasoning sound?

II

RANDOM VARIABLES

1. INTRODUCTION

We have already observed that, in a chance experiment, it is often not the actual outcome that concerns us but some quantity which depends upon the outcome. For example, when we toss a coin ten times we may wish to discuss the number of heads which turn up rather than the exact outcome of each toss. When a gas is considered to be a collection of molecules moving randomly, it may be the pressure or the temperature which we can measure. For this purpose we introduce the notion of a function defined on a sample space.

DEFINITION. A <u>random variable</u> is a numerical valued function with domain a sample space Ω on which a probability measure has been assigned.

The terminology random variable is customary but unfortunate since a random variable is neither random nor a variable. It is simply a function. In calculus we sometimes write a function as $f(x)$ and sometimes as simply f. Similarly, when we speak of random variables, we shall sometimes write X and sometimes $X(\omega)$. Recall that we can represent a function as a map from the domain to the range. Such a map is shown in Figure 1.

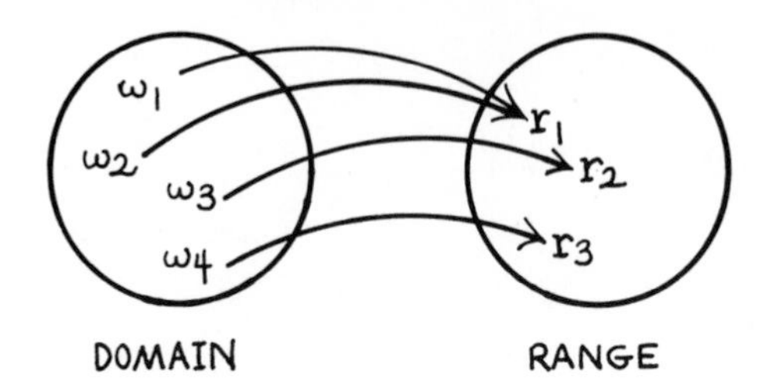

Figure 1

The notation $X(\omega) = a$ means the set of all ω such that $X(\omega) = a$. We shall abbreviate this to $(X(\omega) = a)$ or simply $(X = a)$. Since this set is an event, we can determine its probability, $P(X(\omega) = a)$. This probability will be the sum of the weights assigned to outcomes ω_j such that $X(\omega_j) = a$. The set of all these probabilities as we let a take on all values in the range determines a probability measure on the range R. This measure is called the <u>distribution</u> of X.

DEFINITION. Let X be a random variable defined on a sample space. Let $R = r_1, r_2, \ldots, r_s$ be the range of X. Then the set of probabilities

$$p_j = P(X(\omega) = r_j)$$

for $j = 0,1,2,\ldots,s$ is the <u>distribution</u> of X.

<u>Example 1</u>. A fair coin is tossed three times. Let Ω = {HHH, HHT, HTH, HTT, THH, THT, TTH, TTT}. We assign equal probability to each outcome. Let X be the number of heads which

turn up. Then X has range R = {0,1,2,3}. X(HHH) = 3, X(HHT) = 2, etc. The distribution of X is

$$p(0) = P(X(\omega) = 0) = 1/8$$

$$p(1) = P(X(\omega) = 1) = 3/8$$

$$p(2) = P(X(\omega) = 2) = 3/8$$

$$p(3) = p(X(\omega) = 3) = 1/8$$

We shall often have occasion to consider several random variables defined on the same sample space. For example, in the case of three tosses of a coin, let $X(\omega)$ be the number of heads which turn up, and $Y(\omega)$ be the first time a head turns up. Then X can take on the values 0,1,2,3 and Y the values 1,2,3. For any pair of possible values of X and Y, say r_i, r_j we can compute

$$p(r_i, r_j) = P(X(\omega) = r_i, Y(\omega) = r_j).$$

For example, $p(2,2) = P(X(\omega) = 2,\ Y(\omega) = 2) = P(THH) = 1/8$. The set of all values $p(r_i, r_j)$ is a probability measure on the space of points of the form r_i, r_j where r_i is in the range of X and r_j in the range of Y. This measure is called the joint distribution of X and Y. More generally, we have the following definition:

DEFINITION. If $X_1, X_2, \ldots, X_N$ are random variables defined on a sample space Ω, the <u>joint distribution</u> of these random variables is the set of probabilities of the form

$$P(r_1, r_2, \ldots, r_r) = P(X_1 = r_1, X_2 = r_2, \ldots, X_n = r_n)$$

where r_k is any element of the range of X_k.

We have a concept of independent random variables analogous to that of independent events.

DEFINITION. Random variables $X_1, X_2, \ldots, X_n$ are <u>mutually independent</u> if

$$P(X_1 = r_1, X_2 = r_2, \ldots, X_n = r_n) =$$
$$P(X_1 = r_1)P(X_2 = r_2) \ldots P(X_n = r_n)$$

for any choice of $r_1, r_2, \ldots, r_n$.

We note that in the case of mutually independent random variables, the joint distribution is the product of the individual distributions. When two random variables are mutually independent, we shall say more briefly that they are <u>independent</u>.

<u>Example 1 (cont.)</u>. The random variables X = number of heads, and Y = first time a head turns up, are not independent since $P(X = 1, Y = 2) = 1/8$ but $P(X = 1) = 3/8$ and $P(Y = 2) = 1/4$. Thus $P(X = 1, Y = 2) \neq P(X = 1)P(Y = 2)$. On the other hand, let $X_j = 1$ if the jth outcome is heads and 0 if it is tails. Then each X_j has the same distribution

$$p(0) = P(X_j(\omega) = 0) = 1/2$$

$$p(1) = P(X_j(\omega) = 1) = 1/2$$

The probability of any three values of these random variables occurring, for example, $P(X_1 = 1, X_2 = 0, X_3 = 1)$, is 1/8 and so they are mutually independent.

The study of random variables proceeds by considering special classes of random variables. The first such class that we shall study is the class of <u>independent trials</u>.

DEFINITION. A sequence of random variables $X_1, X_2, \ldots, X_n$ which are mutually independent and which have the same distribution is called an <u>independent trials process</u>. If each experiment has only two possible outcomes we call the process a <u>Bernoulli trials</u> process.

When dealing with a Bernoulli trials process it is customary to label one outcome "success" and the other "failure". The process is then completely determined by giving the probability p for success on any one trial. The probability for failure is then $q = 1-p$.

<u>Example 2</u>. A baseball player, Mickey Mantle, has a batting average of 300. In a typical game he comes to bat three times. If we treat this as a Bernoulli trials process, we assume that each time at bat Mantle gets a hit, "success," with probability .3. The appropriate sample space and probability measure for three times at bat is given by the tree measure in Figure 2.

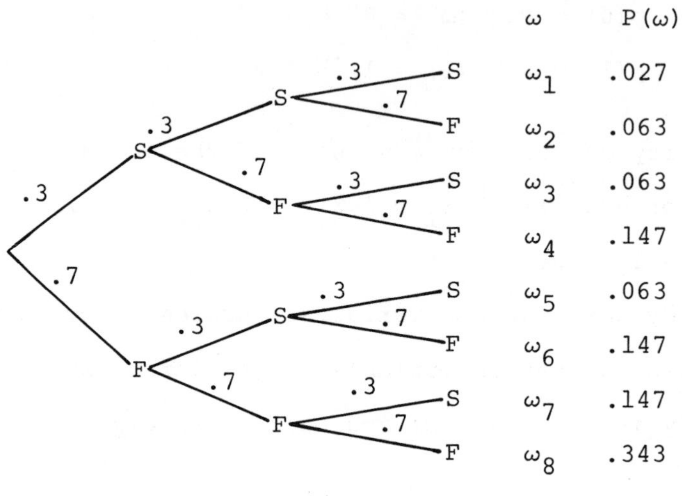

Figure 2.

A particularly interesting random variable is the number of hits that Mantle gets. Let $X(\omega)$ be this number. Then, for example, $X(\omega_2) = 2$, $X(\omega_4) = 1$, etc. The distribution of X is given by

$$P(X(\omega) = 0) = (.7)^3 = .343$$

$$P(X(\omega) = 1) = 3(.3)(.7)^2 = .441$$

$$P(X(\omega) = 2) = 3(.3)^2(.7) = .189$$

$$P(X(\omega) = 3) = (.3)^3 = .027$$

Consider the general case of n times at bat. We cannot draw the tree for n times at bat, but it is easy to visualize the appropriate tree and tree measure. A typical path in the tree is described by a sequence

$$\omega = (S,S,F,F,S,\ldots,S)$$

The probability that we would assign to such a path is

$$P(\omega) = (.3)^j(.7)^{n-j}$$

where j is the number of S's in ω. Let S_n be the number of hits that Mantle gets. $S_n(\omega) = j$ on any path with exactly j S's and n-j F's. The number of such paths is the number of ways that we can choose the j times out of n that Mantle should get a hit. That is, $C(n,j)$. Thus the distribution of S_n is given by

$$P(S_n(\omega) = j) = C(n,j)(.3)^j(.7)^{n-j}$$

for $j = 0,1,2,..,n$.

In the previous example we have found the distribution for the number of successes in a particular case of Bernoulli trials for which $p = .3$. The same reasoning that we used in this example to find this distribution yields the following general result.

THEOREM. Let $S_n(\omega)$ be the number of successes in n Bernoulli trials with probability p for success on each trial. Then

$$P(S_n(\omega) = j) = C(n,j)p^jq^{n-j}$$

for $j = 0,1,2,\ldots,n$

The distribution of S_n is called the binomial distribution and denoted by

$$B(n,j;p) = C(n,j)p^j q^{n-j}.$$

While we have a formula for the binomial distribution, this formula involves the binomial coefficients and as we have seen it is often easier to compute these by an algorithm. The same is true for the binomial distribution. There is only one way to have n successes and hence

$$(1)\quad B(n,n;p) = p^n$$

and similarly

$$(2)\quad B(n,0;p) = (1-p)^n.$$

To obtain j successes in n experiments we must have had j-1 or j successes in the first n-1 experiments. In the first case we would need to have a success on the nth experiment and in the second case a failure on the nth experiment. Thus

$$(3)\quad B(n,j;p) = pB(n-1,j-1;p) + qB(n-1,j;p)$$

for $1 \le j \le n-1$. The relations (1), (2), and (3) determine B(n,j;p) and enable us to modify the program PASCAL to obtain the binomial distribution for any n.

```
BINOMIAL

100 DEF FNF(X) = INT(1000*X+.5)/1000
110 DIM B(50,50)
120 READ P,N
130 FOR M = 0 TO 50
140    LET B(M,0) = (1-P)↑M
150    LET B(M,M) = P↑M
160    FOR J = 1 TO M-1
170       LET B(M,J) = P*B(M-1,J-1) + (1-P)*B(M-1,J)
180    NEXT J
190 NEXT M
200 FOR I = 0 TO N
210    PRINT I, FNF(B(N,I))
220 NEXT I
230 DATA .2,10
240 END
```

```
BINOMIAL

0              0.107
1              0.268
2              0.302
3              0.201
4              0.088
5              0.026
6              0.006
7              0.001
8              0
9              0
10             0
```

We have printed B(10,j;.2). That is, the distribution for 10 independent trials with probability .2 for success on each trial. We note that .2 is the most likely value for the proportion of successes. We expect that p should be the proportion of times the event would happen in the long run. We see that 10 experiments give some evidence of this, but we shall of course have to examine the case of a large number of experiments. This will be the aim of our work in the next sections.

Example 3. Peter and Paul match pennies. By this we shall mean that a coin is tossed a sequence of times. Each time a head turns up, Peter wins 1 dollar and, if tails turns up, he loses a dollar. Let $Z_1, Z_2, \ldots$ be Peter's fortune. If the outcome of the first four tosses is $\omega = (H,H,T,H)$ then $Z_1(\omega) = 1, Z_2(\omega) = 2, Z_3(\omega) = 1$, and $Z_4(\omega) = 2$. The appropriate probability measure for Ω is the independent trials measure with $p = 1/2$. If we call a win for Peter a success and a loss a failure, then we can also define the sequence $S_1, S_2, \ldots$ of successes for Peter. We know that S_n has a binomial distribution $B(n,j;.5)$. We can use this to find the distribution of Z_n. We note that

$$Z_n = S_n - (n-S_n) = 2S_n - n,$$

or

$$S_n = \frac{Z_n + n}{2} .$$

Thus

$$P(Z_n = j) = P(S_n = \frac{j+n}{2}).$$

For example, in 10 plays

$$P(Z_{10} = 8) = P(S_{10} = 9) = B(10,9;.5) = .011.$$

The most likely value for Z_{10} is 0 corresponding to the fact that the distribution of S_{10} has its maximum value at 5.

We shall see that it is typical that when we look at the sum of independent random variables we can expect to have the probabilities increase to a maximum and then decrease. However, we can look at different random variables than the sum associated with the same experiment and obtain quite a different sort of distribution. We close this section with such an example.

Instead of considering only Peter's final fortune in matching pennies, we shall look at a random variable which describes one aspect of his entire sequence of plays. This will be the random variable which represents the number of times that he is in the lead in n matches. If the players are tied neither is in the lead. However, we shall find it convenient to adopt a convention which makes one of the players in the lead when this happens. We shall assume that if they are even the player who was in the lead last time is considered still to be in the lead.

We recall that the program BINARY enabled us to examine all possible sequences of 0's and 1's of length n. That is, all possible plays of the penny matching game. Thus we can easily modify the program to keep track of the number of times Peter is in the lead for any one possible sequence. We can then count the number of such sequences for which he is ahead j times. Dividing this number by 2^n would give the probability that he is ahead j times. Doing this for all j we obtain the distribution for this random variable. The program LEAD is such a modification.

```
LEAD

100 DEF FNF(X) = INT(1000*X+.5)/1000
110 DIM H(12)
120 DATA 12
130 READ N
140 FOR A = 0 TO 2↑N-1
150    LET M = A
160    LET S1 = S = U = 0
170    FOR J = 1 TO N
180       LET R = INT(M/2)
190       LET D = M-2*R
200       LET S = S + 2*D-1
210       IF S < 0 THEN 240
220       IF S1 < 0 THEN 240
230       LET U = U+1
240       LET M = R
250       LET S1 = S
260    NEXT J
270    LET H(U) = H(U) + 1
280 NEXT A
290 FOR U = 0 TO N STEP 2
310    PRINT U,H(U),FNF(H(U)/2↑N)
320 NEXT U
330 END
```

We run the program for n = 12 and print out the number and proportion of paths for which Peter is in the lead 0,2,4,..,12 times. (He must be in the lead an even number of times in 12 tosses.)

```
LEAD

2               504             0.123
4               420             0.103
6               400             0.098
8               420             0.103
10              504             0.123
12              924             0.226
```

We note that the most likely number of times that he is in the lead is either none or all of the time. The central value of

5 is the least likely. This is rather surprising and in complete contrast with the distribution for the fortune S_n. We recall that the distribution of S_n had a bell shaped curve, increasing to a maximum value and then decreasing. We have graphed the distribution of S_{12} in Figure 3.

```
0   *
1   *
2   *
3    *
4          *
5               *
6                 *
7               *
8          *
9    *
10 *
11 *
12 *
```

Figure 3.

We next modify LEAD to print out the graph of the distribution for the number of times Peter is in the lead in 12 matches. We obtain the graph of Figure 4.

```
0                         *
2                *
4              *
6             *
8              *
10                *
12                         *
```

Figure 4.

We see that the graph of the distribution for the number of times in the lead is very different from that we obtained for the fortune of one of the players after 12 plays.

For another example of an interesting random variable associated with penny matching see Exercise 16.

EXERCISES

1. Peter and Paul are matching pennies as in Example 3. Let S_7 be Peter's fortune after 7 tosses. What is $S_7(\omega)$ for:

   (a) ω = (S,S,F,F,F,S,F)

   (b) ω = (S,S,S,F,S,S,F)

   (c) ω = (F,F,S,F,S,S,F)

   (d) ω = (S,F,S,F,F,S,F)

   Ans. -1,3,-1,-1.

2. Joe is playing the following game: a die is rolled a a sequence of times. On each roll, Joe wins a dollar if either a 5 or a 6 turns up and loses a dollar otherwise. Let S_n be Joe's fortune after n plays. What is $S_6(\omega)$ for the following outcome sequences?

   a. ω = 1,3,6,4,1,3

   b. ω = 6,6,2,1,5,2,2

   c. ω = 3,2,5,5,5,3

   d. ω = 1,4,4,5,4,6,3,1

3. In the game of Exercise 2, Joe wins 2 dollars for each time a 5 or 6 turns up and loses 1 dollar otherwise. What is his fortune after each of the following outcome sequences occurs?

   a. ω = 1,1,6,5,2,3

   b. ω = 5,5,6,4,1,6,3

c. $\omega = 4,5,4,1,6,3$

d. $\omega = 2,5,1,3,4,4$

Ans. 0,5,0,-3.

4. Four men, A, B, C, and D, check their hats and they are returned in a random manner. Let Ω be the set of all possible permutations of A,B,C,D. Let $X_j(\omega) = 1$ if the jth man gets his own hat and 0 otherwise. What is the distribution of $X_j(\omega)$? Are $X_1(\omega), X_2(\omega), X_3(\omega), X_4(\omega)$ mutually independent?

Ans. $p_0 = 3/4$, $p_1 = 1/4$, not independent.

5. A box has numbers from 1 to 10. A number is drawn at random. Let X_1 be the number drawn. This number is replaced, and the ten numbers mixed. A second number X_2 is drawn. Find the distribution of X_1 and X_2. Are X_1 and X_2 independent? Answer the same questions for the case where the first number is not replaced before the second is drawn.

6. A die is thrown twice. Let X_1 and X_2 be the outcomes. Define $X(\omega) = \min(X_1(\omega), X_2(\omega))$. For example, if the first outcome is 1 and the second outcome is 3, $\omega = (1,3)$ and $X(\omega) = \min(1,3) = 1$. Find the distribution of X.

Ans. $p_1=11/36, p_2=9/36, p_3=7/36, p_4=5/36, p_5=3/36, p_6=1/36$

7. Given $P(X = a)$, $P(\max(X,Y) = a)$, $P(\min(X,Y) = a)$, determine $P(Y = a)$.

8. Peter and Paul are matching pennies with payoff one dollar as in Example 3. Let S_n denote Peter's fortune after n

matches. At the same time Mary and Nancy are matching pennies. Let T_n denote Mary's fortune. Find $P(T_{1001} > S_{1000})$.

9. Write a program like BINOMIAL and save this for future use in working problems relating to independent trials.

10. A batter has a probability of .3 of getting a hit each time he comes to bat. What is the probability in 50 times at bat that his average will be less than .250?

11. Charles claims that he can distinguish between beer and ale 75 percent of the time. Jim bets that Charles cannot and in fact just guesses. To settle the bet the following test is devised. Charles is asked to try 10 times to identify correctly two glasses, one containing beer and the other ale. If he gets 7 or more correct he wins the bet. What is the probability that Jim wins if Charles is guessing? what is the probability that Charles wins if he does have the ability he claims?

    Ans. .828,.776.

12. In 30 rolls of a die find the probability that a six turns up exactly 5 times. What is the most probable number of times that a six will turn up?

13. Modify the program BINOMIAL to print a graph of the distribution $B(n,j;p)$ for $j = 0,1,2,...,n$ and fixed n and p. Run the program for $n = 20$ and $p = .2$.

14. Show that $B(n,j;p) = p(n-j+1)/(jq)B(n,j-1;p)$

15. For given n, p find the value of j which makes $B(n,j;p)$ the greatest. (Hint: Consider the successive ratios as j increases and use Exercise 14.)

16. Modify the program LEAD to find the distribution of the maximum value of Peter's fortune in 10 penny matches. Compare this distribution with that obtained by the program LEAD.

17. Smith and Jones play a game in which Smith must win a points before Jones wins b points. Assume that Smith wins each game with probability p and Jones with probability $q = 1-p$. Let $P(m,n)$ be the probability that Smith wins when he has won m points and Jones has won n points. Show that

    (a) $P(a,n) = 1 \quad n < b$

    $P(m,b) = 0 \quad m < a$

    (b) $P(m,n) = pP(m+1,n) + qP(m,n+1) \quad m < a,\ n < b.$

    Show that (a) and (b) determine $P(m,n)$ for all $m \leq a$ and $n \leq b$. Write a program to compute $P(m,n)$ for given m,n,a,b, and p.

18. The Yankees are playing the Dodgers in a world series. the Yankees win each game with probability .6. Using your program in Exercise 17 find the probability that they win the series. That is, they win four games before the Dodgers win four games.

    Ans. .71

19. Assume that the random variables X and Y have the joint

probability distribution given by

		Y			
		-1	0	1	2
	-1	0	1/36	1/6	1/12
X	0	1/18	0	1/18	0
	1	0	1/36	1/6	1/12
	2	1/12	0	1/12	1/6

(a) What is $P(X \geq 1 \text{ and } Y \leq 0)$?

(b) What is the conditional probability that $Y \leq 0$ given that $X = 2$?

(c) Are X and Y independent?

(d) What is the probability distribution of $Z = XY$?

2. EXPECTED VALUE

When a large collection of numbers is assembled as in the census, we are not presented with all the numbers but rather certain descriptive quantities such as the average or the median. It is often convenient to do the same for the distribution of a random variable. In this and the next section we shall discuss two such quantities. The first is the expected value. To give some intuitive justification for our definition we consider the following game:

A die is rolled. If an odd number turns up, we win an amount corresponding to the outcome on the die. If an even number turns up we lose an amount equal to the outcome. For example, if a 2 turns up we lose 2 and if a 3 comes up we win 3. We want to decide if this is a reasonable game to play. We shall first simply try it out by simulation.

The program DIE carries out this simulation. We have chosen the data to simulate rolling a die a sequence of times. The program also prints out the amount we would win in each roll and the frequency and the relative frequency that each outcome occurs. Finally it prints the average winnings. We have run the program 3 times. In the first two runs we have played the game 100 times. In the first run our average gain is -.41. In the second it is -.09. It looks like the game is unfavorable and it is natural to ask how unfavorable it really is. To get a better idea we have made 100,000 plays in our third run. (We omit individual outcomes.) In this case our average gain is -.504.

DIE

```
100 RANDOMIZE
110 READ R                                    'NO. OUTCOMES
120 FOR I = 1 TO R
130    READ X(I),P(I)                         'X IS OUTCOME, P IS PROB.
140 NEXT I
150 READ N                                    'N IS NO. TRIALS
160 PRINT
170 FOR K = 1 TO N
180    LET T = RND
190    FOR J = 1 TO R
200       LET T = T - P(J)
210       IF T<0 THEN 240
220    NEXT J
230    PRINT
240    LET F(J) = F(J) + 1
250    PRINT X(J);
260 NEXT K
270 PRINT
290 PRINT "X";TAB(10);"FREQ.";TAB(20);"REL. FREQ."
300 FOR J = 1 TO R
310    PRINT X(J);TAB(11);F(J);TAB(21);F(J)/N
320    LET S = S + X(J)*F(J)/N
330 NEXT J
340 PRINT
350 PRINT "AVERAGE WINNINGS = ";S
360 DATA 6
370 DATA 1,.16667,-2,.16667,3,.16667,-4,.16667,5,.16667,-6,.16667
380 DATA 100
390 END
```

DIE

```
-6 -6  3  3  3  1  1  5  5 -6 -6  5 -6 -2  1  3  3 -6 -6  1  5  3
-4  3  1  3 -6  5  3  3 -6  1 -2 -4  1  1 -2  1 -6 -2 -4  1  1 -4
-4 -6 -2  1  5  3  5 -4 -4  5  1  5 -6  3  3 -4 -6 -2 -4 -2 -2  1
-2  1  5 -6  3  1  5 -6  3 -4  1 -4  3 -6  5 -6  1  1 -4 -2  5 -4
-6 -6  3 -2  5 -4  3 -6  1  3  5  5
```

X	FREQ.	REL. FREQ.
1	20	0.2
-2	11	0.11
3	19	0.19
-4	14	0.14
5	16	0.16
-6	20	0.2

AVERAGE WINNINGS = -0.41

DIE

```
1 -2  3  3 -2  5  3 -2 -4  3  1  3 -4 -2 -4  3 -4  5 -6  3 -2  5
5  1 -6  5 -4 -6 -4 -2  3  5  5 -4  1  3  1  1  1  5  5 -6 -4  3
5 -6  1  1 -6  5  5  5  1  3 -6  3  5 -4 -6 -2  3 -6  3  3 -4  1
5  1  5 -6 -4 -6 -6  5 -6 -6 -6  3 -4 -4  5  1  5  5 -2 -6  1 -2
5 -4 -6  5 -2  1 -4 -2 -4  3 -2  5
```

X	FREQ.	REL. FREQ.
1	15	0.15
-2	12	0.12
3	17	0.17
-4	16	0.16
5	23	0.23
-6	17	0.17

AVERAGE WINNINGS = -0.09

DIE

X	FREQ.	REL. FREQ.
1	16590	0.1659
-2	16671	0.16671
3	16634	0.16634
-4	16640	0.1664
5	16706	0.16706
-6	16759	0.16759

AVERAGE WINNINGS = -0.50434

We note that the relative frequency of each of the 6 possible outcomes is quite close to the probability 1/6 for a particular outcome on one roll This corresponds to our frequency interpretation of probability. It also suggests that for very large numbers of plays our average winning should be

$$\mu = 1(1/6)-2(1/6)+3(1/6)-4(1/6)+5(1/6)-6(1/6)$$

$$= 9/6 - 12/6 = -3/6 = -.5.$$

This agrees quite well with our average gain for 100,000 plays.

We note that the value we have chosen for our estimate for the average winning is obtained by taking the possible outcomes, multiplying by the probability, and adding the results. This suggests the following more general definition for the expected outcome of an experiment.

DEFINITION. Let X be a random variable defined on a sample space Ω with probability measure P. The <u>expected</u> <u>value</u> E(X) is defined by

$$E(X) = \sum_{\omega} X(\omega)P(\omega).$$

We sometimes write $\mu = E(X)$.

Of course, just as with the frequency definition of probability, we shall need to make this more precise. We know that for any finite experiment the average of the outcomes is not predictable. However we shall show that the average will be close to E(X) for a large number of experiments. We shall first need to develop some properties of the expected value. Using these properties and the concept of the variance to be introduced in the next section, we shall be able to prove the law of large numbers. This theorem will justify mathematically both our frequency concept of probability and the interpretation of "expected value" as the average value to be expected in a large number of experiments.

From the definition of expected value we see that if X and Y are two such random variables relative to Ω

$$
\begin{aligned}
E(X + Y) &= \sum_{\omega} (X(\omega) + Y(\omega))P(\omega) \\
&= \sum_{\omega} X(\omega)P(\omega) + \sum_{\omega} Y(\omega)P(\omega) \\
&= E(X) + E(Y).
\end{aligned}
$$

Also, if a is any constant

$$
\begin{aligned}
E(aX) &= \sum_{\omega} aX(\omega)P(\omega) \\
&= a\sum_{\omega} X(\omega)P(\omega) \\
&= aE(X).
\end{aligned}
$$

Combining these two results we have the following theorem.

THEOREM. If X and Y are any two random variables and a and b are any two numbers then

$$E(aX + bY) = aE(X) + bE(Y).$$

We next show that to compute the expected value of a random variable it is enough to know its distribution. Let X be a random variable with range $r_1, r_2, \ldots, r_s$ and probability distribution $p_1, p_2, \ldots, p_s$. Let

$$A_j = \{\omega : X(\omega) = r_j\}.$$

Then $p_j = P(A_j)$. Thus

$$\begin{aligned} E(X) &= \sum_{\omega} X(\omega)P(\omega) \\ &= \sum_j \sum_{\omega \text{ in } A_j} X(\omega)P(\omega) \\ &= \sum_j r_j P(X(\omega) = r_j) \\ &= \sum_j r_j p_j. \end{aligned}$$

Example 1. A fair coin is tossed three times. Let $X_j = 1$ if the jth outcome is heads and 0 if it is tails, for $j = 1,2,3$. Then $S_3 = X_1 + X_2 + X_3$ gives the number of heads in three tosses. We know that S_3 has a binomial distribution $p_0 = 1/8, p_1 = 3/8, p_2 = 3/8, p_3 = 1/8$. Thus,

$$\begin{aligned} E(S_3) &= 0(1/8) + 1(3/8) + 2(3/8) + 3(1/8) \\ &= 12/8 = 3/2. \end{aligned}$$

A second way to compute this expected value is to use our theorem. We first find

$$E(X_j) = 0(1/2) + 1(1/2) = 1/2.$$

Then

$$\begin{aligned} E(S_3) &= E(X_1) + E(X_2) + E(X_3) \\ &= 3/2. \end{aligned}$$

This example is a special case of Bernoulli trials. Consider the general case of Bernoulli trials with probability p for success and $q = 1-p$ for failure. Let $X_j(\omega) = 1$ if the jth outcome is a success and 0 if it is a failure. Then

$$E(X_j) = 1p + 0q = p.$$

Then if $S_n = X_1+X_2+ \ldots + X_n$, we know that S_n is the number of successes and

$$\begin{aligned} E(S_n) &= E(X_1) + E(X_2) + \ldots + E(X_n) \\ &= np \end{aligned}$$

If X and Y are two random variables, it is not true in general that $E(XY) = E(X)E(Y)$. However, this is true if X and Y are independent.

THEOREM. If X and Y are independent random variables, then

$$E(XY) = E(X)E(Y)$$

Proof:

$$E(XY) = \sum_{\omega} X(\omega)Y(\omega)P(\omega)$$

$$= \sum_{j,k} r_j r_k P(X(\omega) = r_j, Y(\omega) = r_k)$$

But if X and Y are independent

$$P(X(\omega) = r_j, Y(\omega) = r_k) = P(X(\omega) = r_j)P(Y(\omega) = r_k).$$

Thus

$$\begin{aligned} E(XY) &= \sum_{j,k} r_j r_k P(X(\omega) = r_j)P(Y(\omega) = r_k) \\ &= (\sum_j r_j P(X(\omega) = r_j))(\sum_k r_k P(Y(\omega) = r_k)) \\ &= E(X)E(Y). \end{aligned}$$

Example 1. cont.. We know that X_1 and X_2 are independent. They each have expected value 1/2. Thus $E(X_1)E(X_2) = (1/2)(1/2) = 1/4$.

Example 2. We next give a simple example to show that the expected values need not multiply if the random variables are not independent. To see this we consider a single toss of a coin. We define the random variable X to be 1 if heads turns up and 0 if tails turns up. We define Y to be 0 if heads turns up and 1 if tails turns up. Then $E(X) = E(Y) = 1/2$. But $XY = 0$ for either outcome. Hence $E(XY) = 0$.

The concept of expectation plays an important role in analyzing gambling games. The problems mentioned in the discussion of the origins of probability were often asked by a gambler wishing to evaluate his chances in a gambling game. See, for example, Exercise 15. The simplest kinds of gambling games can be considered to be independent trials processes. For example, playing a sequence of games of roulette making the same kind of bet each time. For such games we make the following definition:

DEFINITION. A player plays a game in which his winnings on successive plays is an independent trials process $X_1, X_2, \ldots$. We say that the game is unfavorable if $E(X_j) < 0$, fair if $E(X_j) = 0$, and favorable if $E(X_j) > 0$.

We recall that each random variable in an independent trials process has the same distribution and hence the same expected value. Thus to determine the fairness of a game we need only consider the outcome of a single play. For example, in the case of penny matching, each play results in +1 with probability 1/2 or -1 with probability 1/2. Thus the game is fair since the expected winning on any one play is 0. This is a special case of a chance process called a martingale.

If the winnings of a player in a game are given by an independent trials process $X_1, X_2, \ldots$. then his fortune after n plays is given by

$$S_n = X_1 + X_2 + \ldots + X_n.$$

Example 3. Let us consider the familiar gambling game of craps. The simplest version of this game is played as follows. A player rolls a pair of dice. If the sum of the points which turn up is

7 or 11 he wins one dollar. If it is 2,3, or 12 he loses one dollar. In any other case he continues to roll the dice until he either gets a sum equal to that on the first roll or gets a sum of 7. In the first case he wins one dollar and in the latter case he loses one dollar. Let us compute his expected winning on any one play to decide if the game is unfavorable, fair, or favorable. To determine this we must analyze the possible outcomes of a single play of the game. To do this we construct a two stage tree measure as shown in Figure 5.

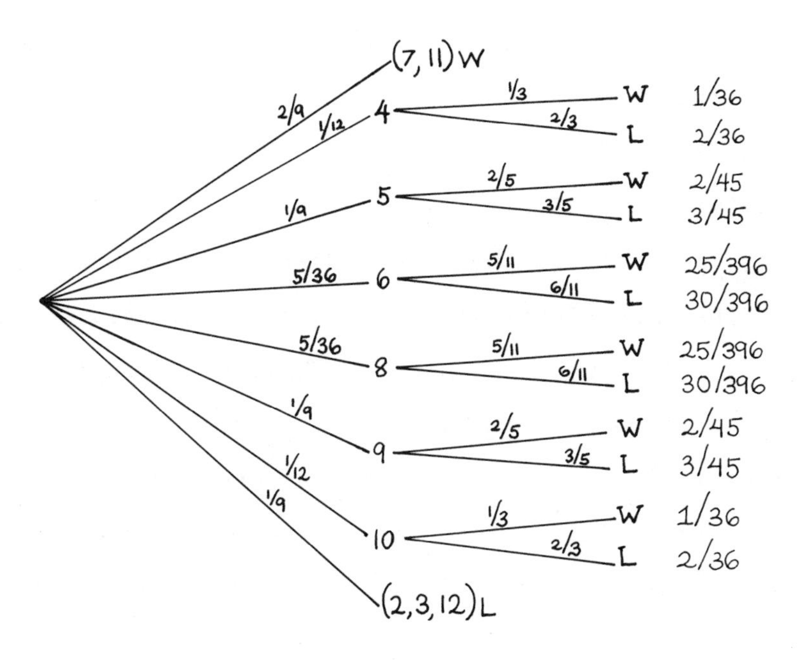

Figure 5.

The first stage represents the possible sums for his first roll. The second stage represents the possible outcomes for the game if it is not ended on the first roll. In this stage we are representing the possible outcomes of a sequence of rolls required to determine the final outcome. The branch probabilities for the first stage are computed in the usual way assuming all 36 possibilities for outcomes for the pair of die to be equally likely. For the second stage we assume that the game will eventually end and compute conditional probabilities given that it does eventually end. For example, assume that the player obtained a 6 on the first roll. Then the game will end when one of the eleven different pairs

(1,5),(2,4),(3,3),(4,2),(5,1)

(1,6),(2,5),(3,4),(4,3),(5,2),(6,1)

occurs. We assume that each of these possible pairs has the same probability. Then the player wins in the first 5 cases and loses in the last 6. Thus we take his probability of winning to be 5/11 and losing to be 6/11. From the path probabilities we can find the probability that the player wins one dollar. This is 244/495. The probability that he loses one dollar is then 251/495. Thus if X_j is his winning on the jth play,

$$\begin{aligned} E(X_j) &= 1(244/495) + (-1)(251/495) \\ &= -7/495 = -.0141 \end{aligned}$$

Thus the game is unfavorable, but only slightly. His expected loss in n plays is -n(.0141). If n is not large this is a small average loss for the player. The gambling house, of course, assumes a large number of plays taking into account all the players that play the game in a day and so can afford a small average winning per play and still make large sums.

It is a favorite pastime to develop systems of play for gambling games and for other games such as the stock market. One of the important properties of a fair game is that the game cannot be beaten in terms of expectation by various systems of play. We close this section with a simple illustration of this fact.

Example 4. Let us assume that a stock increases or decreases each day in value by one dollar with equal probability. Then for this simplified model we have our familiar matching pennies game. We assume that a buyer, Mr. Ace, adopts the following strategy. He buys the stock on the first day at its price V. He then waits until the price of the stock increases by one to $V+1$ and sells. He then continues to watch the stock until its price falls back to V. He buys again and waits until it goes up to $V+1$ and sells. Thus he holds the stock in intervals during which it increases by one dollar. In each such interval he makes a profit of 1. However, we assume that he can do this only for a finite number of trading days. Thus he can lose if in the last interval that he holds the stock it does not get back up to $V+1$. This is the only way he can lose. In Figure 6 we illustrate a typical history if Mr. Ace must stop in ten days.

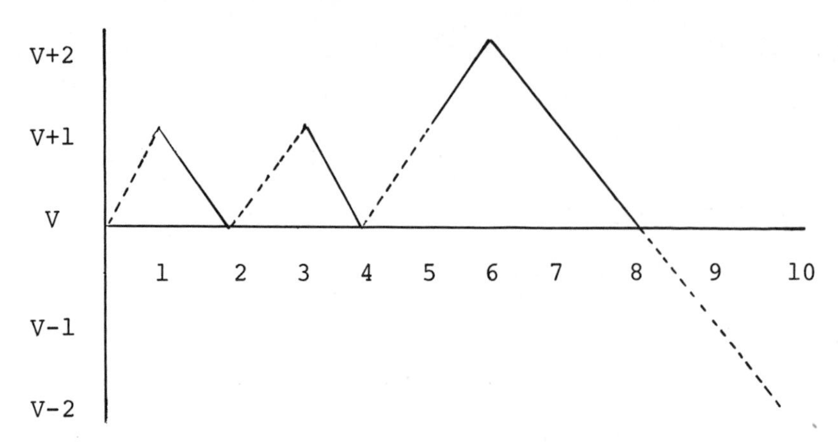

Figure 6.

Mr. Ace holds the stock under his system during the days indicated by broken lines. We note that his system nets him a gain of 3 during the periods when his system is successful but he loses 2 during his last holding since it does not get back up to 1 before he must quit.

Since our model for the increases of the stock price is that of the penny matching experiment, it is again an easy matter to use our basic program BINARY to keep track for every possible sequence of stock gains over an N day period the fortune of Mr. Ace using his system. The program SYSTEM is the appropriate modification and a run of the program gives the distribution of his gains and the expected value of the gain using his system over a ten day period. The price of the stock is denoted by S. S1 represents the price of the stock during the periods when Mr. Ace holds a share of the stock. If the test in line 200 and in line 210 are both passed, then the price has gone from 0 to 1 during this period and we increase his profit T by one in line 220. We note from a run of the program that his expected profit is 0.

```
SYSTEM

100 DEF FNF(X) = INT(1000*X+.5)/1000
110 DIM H(100)
120 READ N
130 DATA 10
140 FOR I = 0 TO 2↑N-1
150    LET M = I
160    LET S1 = S = T = 0
170    FOR J = 1 TO N
180       LET R = INT(M/2)
182       LET C = M-2*R
185           LET M = R
190       LET S = S + 2*C-1
200       IF S1 <> 0 THEN 230
210       IF S <> 1 THEN 230
220       LET T = T + 1
230       LET S1 = S
240    NEXT J
250    IF S > 0 THEN 270
260    LET T = T + S
270    LET H(T+N) = H(T+N) + 1
280 NEXT I
290 FOR L = 0 TO N+N/2
300    LET U = (L-N)*H(L) + U
310    PRINT L-N,H(L),FNF(H(L)/2↑N)
320 NEXT L
330 PRINT "EXPECTED WINNING UNDER SYSTEM =";U/2↑N
325 PRINT
400 END
```

SYSTEM

-10	1	.001
-9	0	0
-8	9	.009
-7	1	.001
-6	35	.034
-5	9	.009
-4	76	.074
-3	35	.034
-2	99	.097
-1	76	.074
0	77	.075
1	309	.302
2	196	.191
3	80	.078
4	19	.019
5	2	.002

EXPECTED WINNING UNDER SYSTEM = 0

While the expected value remains 0 we note that the probability that Mr. Ace is ahead after ten days is .592. If he did not use a system this probability would be slightly less than 1/2. Thus he would be able to tell his friends that his system gave him a better chance of being ahead of the game than someone who simply buys the stock and holds it,assuming, of course that our simple random model is correct. There have been a number of studies to determine how random the stock market is. At least one Congressman achieved public attention by claiming that he did better than the mutual funds by throwing darts at a dart board to determine which stocks to buy and sell.

EXERCISES

1. A card is drawn at random from a deck consisting of cards numbered 2 through 10. A player wins one dollar if the

number on the card is odd and loses one dollar if the number is even. What is the expected value of his winning?

Ans. -1/9

2. A card is drawn at random from a deck of cards. If it is red a player wins one dollar. If it is black he loses $2. What is the expected value of the game?

3. In a class there are 20 students, 6 are age 18, 10 are age 19, and 4 are age 20. A student is chosen at random. What is his expected age?

Ans. 18.9

4. One version of roulette at Monte Carlo is played as follows: A wheel has the number 0 and the numbers 1 to 36 marked on equal slots. The wheel is spun and a ball stops randomly at one slot. If a player puts one dollar on a number, he receives $36 if the ball stops on that number. His net win is $35. Otherwise he loses his dollar. Find the expected value for his net winning.

Ans. -1/37 = -.027.

5. In a second version of roulette, a player may bet on red or black. Half of the numbers from 1 to 36 are red and half are black. If a player bets on black and a black number turns up he gets his dollar back and another dollar. If a red number turns up he loses his dollar. If the ball stops on 0, play continues until the ball stops on a color. If it is red, the player receives his original dollar back, but no

more. Otherwise he loses his dollar. Find the expected value of the game.

6. Write a program to simulate roulette as played in Exercise 4. Run the program for 1000 plays and find your average winning. Do the same for the version of roulette described in Exercise 5.

7. Let X denote the sum of the values showing on a toss of two fair dice, and Y the difference of the values. Show that $E(XY) = E(X)E(Y)$. Are X and Y independent?

8. Show that if X and Y are random variables taking on only two values each, and if $E(XY) = E(X)E(Y)$, then X and Y are independent.

9. A royal family has children until they have a boy or until they have had three children. Assume that each child is a boy with probability 1/2. Find the expected number of boys in this royal family and the expected number of girls.

10. Write a computer program to simulate the game of craps. Play the game 1000 times and compare your average winning with the expected winning on a single play.

11. In Example 4, assume that Mr. Ace decides to buy the stock and to hold it until it goes up one dollar and then sell and not buy again. Modify the program SYSTEM to find the distribution of his profit under this system after a ten day period. Find the expected profit and the probability that he comes out ahead.

12.* A man carries in each of his two front pockets a box of matches originally containing N matches. Whenever he need a match he chooses a pocket at random and removes one from that box. One day he reaches into a pocket and finds the box empty.

a. Let p_r denote the probability that the other pocket contains r matches. Define a sequence of counter random variables as follows. Let $X_i = 1$ if the i'th draw is from the left pocket and 0 if it is from the right pocket. Interpret p_r in terms of $S_n = X_1 + X_2 + ... + X_n$. Find a binomial expression for p_r.

b. Write a computer program to compute the p_r, as well as the probability that the other pocket contains at least r matches for N = 100 and r from 0 to 50. c. Show that the p_r satisfy

$$(N-r)p_r = 1/2(2N+1)p_{r+1} - 1/2(r+1)p_{r+1}.$$

d. Evaluate $\sum_r p_r$.

e. Use (c) and (d) to determine the expectation E of the distribution $\{p_r\}$.

f. Use Stirling's formula to obtain an approximation for E. How many matches must each box contain to ensure a value about 13 for the expectation E (take $\pi = 22/7$).

13. In the hat check problem it was assumed that N people check their hats and they are handed back at random. Let $X_j = 1$

if the jth man gets his hat and 0 otherwise. Find $E(X_j)$ and $E(X_j X_k)$ for j not equal to k. Are X_j and X_k independent?

14. Prove by a combinatorial argument that

$$\sum_j jB(n,j;p) = np.$$

Interpret your result in probabilistic terms.

15. Player 1 throws six dice and wins one dollar if at least one ace turns up; player 2 throws 12 dice and wins one dollar if 2 or more aces turn up. Which is the better game?

16. A box contains 2 gold balls and 3 lead balls. You are allowed to choose successively balls from the box at random. You win one dollar each time you draw a gold ball and lose one dollar each time you draw a lead ball. After a draw the ball is not replaced. Find

a. Your expected winning if you draw two balls and quit.

b. Your expected winning if you make five draws, i.e., draw all of the balls out.

c. Your expected winning if you continue to draw until either you are ahead by one dollar or there are no more balls.

3. VARIANCE

The usefulness of the expected value as a number descriptive of the outcome of an experiment is increased when the outcome is not likely to deviate too much from the expected value. In this section we shall introduce a measure of this deviation called the variance.

DEFINITION. Let X be a random variable with $\mu = E(X)$. Then the <u>variance</u> of X, denoted by $V(X)$, is

$$V(X) = E((X-\mu)^2).$$

and the <u>standard deviation</u> of X, denoted by $D(X)$, is $D(X) = V(X)^{1/2}$.

We often denote $V(X)$ by σ^2 and $D(X)$ by σ.

Note that from the definition of variance we can write $V(X)$ in the form:

$$(1) \quad V(X) = \sum_{\omega} (X(\omega)-\mu)^2 P(\omega)$$

<u>Example 1</u>. Consider the case of one roll of a die. Let X be the number which turns up. To find the $V(X)$ we first find the expected value of X. This is,

$$E(X) = 1(1/6) + 2(1/6) + 3(1/6) + 4(1/6) + 5(1/6) + 6(1/6)$$
$$= 7/2$$

To find the variance of X we form the new random variable $(X-\mu)^2$ and compute its expectation. We can easily do this from the following table.

ω	$P(\omega)$	$X(\omega)$	$(X(\omega) - 7/2)^2$
1	1/6	1	25/4
2	1/6	2	9/4
3	1/6	3	1/4
4	1/6	4	1/4
5	1/6	5	9/4
6	1/6	6	25/4

From this table we find $E(X-\mu)^2$ is

$$V(X) = 1/6(25/4 + 9/4 + 1/4 + 1/4 + 9/4 + 25/4)$$

$$= 35/12$$

The variance has very different properties than the expectation. If c is any constant, $E(cX) = cE(X)$, $E(X+c) = E(X) + c$. In contrast we have the following properties of the variance.

THEOREM. If c is any constant and X any random variable, then

$$V(cX) = c^2V(X)$$

$$V(X+c) = V(x).$$

Proof. Let $\mu = E(X)$. Then $E(cX) = c\mu$. Thus

$$V(cX) = E((cX - c\mu)^2) = E(c^2(X - \mu)^2)$$

$$= c^2E((X-\mu)^2) = c^2V(X).$$

To prove the second assertion we note that to compute $V(X+c)$ we would replace $X(\omega)$ by $X(\omega)+c$ and μ by $\mu+c$ in (1). But the c would cancel out in this case and give us $V(X)$.

We shall next prove a result which gives us a useful alternative form for computing the variance.

THEOREM. If X is any random variable with $E(X) = \mu$, then

$$V(X) = E(X^2) - \mu^2.$$

Proof. We note that

$$V(X) = E(X - \mu)^2 = E(X^2 - 2\mu X + \mu^2)$$
$$= E(X^2) - 2\mu E(X) + \mu^2 = E(X^2) - \mu^2.$$

Using this result we can compute the variance of the outcome of a roll of a die by first computing

$$E(X^2) = 1(1/6)+4(1/6)+9(1/6)+16(16)+25(1/6)+36(1/6)$$
$$= 91/6$$

and hence

$$V(X) = E(X^2) - \mu^2 = 91/6 - (7/2)^2$$

$$= 35/12.$$

We shall find the result just proved useful for computing the variance by hand and for certain proofs. When computing the variance with a computer it is better to go back to the definition. The reason for this is that the numbers that occur in the computation of $E(X-\mu)^2$ are smaller than those in the computation of $E(X^2)$.

Two distributions that we have studied that have the same expectation are the distribution of the number of heads which turn up in 10 penny matching plays and the number of times a particular player is in the lead during these plays. We computed these distributions in the programs BINOM and LEAD. We now plot these distributions and give the expected value,variance, and standard deviations for these two distributions.

```
*                     0          .001
*                     1          .01
  *                   2          .044
     *                3          .117
          *           4          .205
            *         5          .246
          *           6          .205
     *                7          .117
  *                   8          .044
*                     9          .01
*                     10         .001
```

EXPECTED VALUE = 5

VARIANCE = 2.5

STANDARD DEVIATION = 1.581

```
         *         0         .246
     *             2         .137
    *              4         .117
    *              6         .117
     *             8         .137
         *         10        .246
```

EXPECTED VALUE = 5

VARIANCE = 15.

STANDARD DEVIATION= 3.873

We see that at the expected value the binomial distribution has its maximum, and the distribution for the number of leads has its minimum. This results in a significantly smaller variance for the binomial case as we would expect.

We return now to some general properties of the variance. We recall that if X and Y are any two random variables, $E(X+Y) = E(X) + E(Y)$. This is not true generally for the case of the variance. For example, let X be any random variable and define $Y = -X$. Then $V(X) = V(Y)$ so that $V(X) + V(Y) = 2V(X)$. But $X+Y$ is always 0 and hence has variance 0.

We shall now show that the variance of the sum of two independent random variables is the sum of the variances.

THEOREM. Let X and Y be two independent random variables. Then $V(X + Y) = V(X) + V(Y)$.

Proof. Let $E(X) = a$ and $E(Y) = b$. Then

$$V(X + Y) = E((X + Y)^2) - (a + b)^2$$
$$= E(X^2) + 2E(XY) + E(Y^2)-a^2-2ab-b^2.$$

Since X and Y are independent $E(XY) = E(X)E(Y) = ab$. Thus

$$V(X+Y) = E(X^2)-a^2 + E(Y^2)-b^2 = V(X) + V(Y).$$

It is easy to see that this result extends to any number of random variables.

We can thus apply this result to independent trials processes to obtain the following result.

THEOREM. Let X_1, X_2, ..., be an independent trials process. Let

$$S_n = X_1 + X_2 + \dots + X_n$$

and

$$A_n = S_n/n.$$

Then

$$E(S_n) = n\mu, \; V(S_n) = n\sigma^2$$
$$E(A_n) = \mu, \; V(A_n) = \sigma^2/n.$$

Proof. Since all the random variables X_j have the same expected value we have

$$E(S_n) = E(X_1) + \quad + E(X_n) = n\mu$$

$$V(S_n) = V(X_1) + \dots + V(X_n) = n\sigma^2.$$

We have seen that if we multiply a random variable X with mean μ and variance σ^2 by a constant c, the new random variable has expected value $c\mu$ and variance $c^2\sigma^2$. Thus $E(A_n) = E(S_n/n) = n\mu/n = \mu$. Also $V(A_n) = V(S_n/n) = V(S_n)/n^2 = n\sigma^2/n^2 = \sigma^2/n$.

Example 2. Consider n rolls of a die. We have seen that if X_j is the outcome of the jth roll then $E(X_j) = 7/2$ and $V(X_j) = 35/12$. Thus if S_n is the sum of the outcomes and $A_n = S_n/n$ is the average of the outcomes, we have that $E(A_n) = 7/2$ and $V(A_n) = (35/12)n$. Thus as n increases the expected value of the average remains constant, but the variance will tend to 0. If the variance is a true measure of the expected value of the deviations from the mean, this should say that for large n we can expect the average to be very near the expected value. This is the case and we shall justify this in the next section.

We consider finally the important special case of Bernoulli trials. We shall as usual let $X_j = 1$ if the jth outcome is success and 0 if it is failure. If p is the probability of a success, and $q = 1-p$,

$$E(X_j) = 0q + 1p = p$$
$$E(X_j^2) = 0^2q + 1^2p = p$$

and

$$V(X_j) = E(X_j^2) - (E(X_j))^2 = p - p^2 = pq$$

Thus $E(A_n) = p$ and $V(A_n) = pq/n$ where A_n is the average number of successes. Again we see that the expected proportion of successes remains p and the variance tends to 0. This again suggests that the frequency interpretation of probability is correct. We make this more precise in the next section.

EXERCISES

1. A number is chosen at random from the set $S = \{-1,0,1\}$. Let X be the number chosen. Find the variance and standard deviation of X.
2. A die is loaded so that the probability of a face coming up is proportional to the number on that face. The die is rolled with outcome X. Find V(X) and D(X).

   Ans. $V(X) = 980/21^2$, $D(X) = (980)^{1/2}/21$.
3. Prove the following facts about the standard deviation.

   (a) $D(X+c) = D(X)$.

   (b) $D(cX) = |c|D(X)$.
4. A coin is thrown three times. Let X be the number of heads which turn up. Find the V(X) and D(X).
5. A number is chosen at random from the integers $1,2,3...,n$. Let X be the number chosen. Show that $E(X) = (n+1)/2$ and $V(X) = (n-1)(n+1)/12$.
6. Prove that $V(X) = 0$ if and only if X is a constant function.
7. Let X be a random variable with $\mu = E(X)$ and $\sigma^2 = V(X)$. Let $X^* = (X - \mu)/\sigma$. The random variable X^* is called the

standardized random variable associated with X. Show that this standardized random variable has expected value 0 and variance 1.

8. Peter and Paul match pennies. Let S_n be Peter's fortune after n matches. Show that $V(S_n) = n$.

9. In the problem of the children of a royal family given in Exercise 9 of the previous section, find the variance for the number of boys and for the number of girls in the family.

10. Once more n people have their hats returned at random. Let $X_i = 1$ if the ith man gets his own hat back and 0 otherwise. Let $S_n = \sum_j X_j$. Then S_n is the total number of people that get their own hats back. Show that

    a. $E(X_i^2) = 1/n$

    b. $E(X_i X_j) = 1/(n(n-1))$ for $i \neq j$.

    c. Using (a) and (b) show that $E(S_n^2) = 2$.

    d. Show that $V(S_n) = 1$ independent of n.

11. A die is rolled twice. Let X be the maximum of the numbers obtained. Find the variance of X.

12. Write a program to find the variance of a random variable and use this program to find the variance for the profit made by Mr. Ace by his system described in the previous section. Compare this with the variance of his profit if he holds the stock for ten days.

13. Let S_n be the number of successes in n independent trials.

Modify the program BINOMIAL to compute, for given n,p and k, the probability

$$P(-kD(S_n) < S_n - n\mu < kD(S_n)).$$

For p = .5 and for each of the values k = 1,2,3, compute this probability for n = 10,30,50. What does the result of your computation suggest for the standardized random variable S_n^*? Do the same for the case p = .2.

III

LIMIT THEOREMS

1. LAW OF LARGE NUMBERS

We are now in a position to prove our first fundamental theorem of probability, which justifies our frequency interpretation of probabilities, and our interpretation of the expected value as an average outcome. This theorem,called the Law of Large Numbers, is the mathematical version of the idea more often expressed as the "law of averages."

To discuss the law of large numbers, we first need a fundamental inequality called the Chebyshev inequality. Let X be any random variable with range R. Let $p(r) = P(X(\omega) = r)$ and $\mu = E(X)$. Let $e > 0$ be any number. Let A be the set of all such that $|X(\omega) - \mu| \geq e$. Then

$$V(X) = \sum_{\omega} (X(\omega) - \mu)^2 P(\omega)$$

$$\geq \sum_{r \text{ in } A} (X(\omega) - \mu)^2 P(\omega)$$

since we are simply leaving out some positive terms. On the other hand $(X(\omega) - \mu)^2 \geq e^2$ for r in A. Thus

$$V(X) \geq e^2 \sum_{r \text{ in } A} P(\omega)$$

$$= e^2 P(|X(\omega) - \mu| \geq e)$$

This gives us the <u>Chebyshev inequality</u>:

$$P(|X(\omega) - \mu| \geq e) \leq \frac{V(X)}{e^2}.$$

Note that X is an arbitrary random variable.

We are now prepared to prove the law of large numbers. Let $X_1, X_2, \ldots$ be an independent trials process with $\mu = E(X_j)$ and $\sigma^2 = V(X_j)$. Then if $S_n = X_1 + X_2 + \ldots + X_n$,

$$V(S_n) = nV(X_1) = n\sigma^2.$$

and

$$V(\frac{S_n}{n}) = \frac{V(S_n)}{n^2} = \frac{n\sigma^2}{n^2} = \frac{\sigma^2}{n}.$$

Also we know that $E(\frac{S_n}{n}) = \mu$.

By Chebyshev's inequality, for any $e > 0$,

$$P(|\frac{S_n}{n} - \mu| \geq e) \leq \frac{\sigma^2}{ne^2}$$

Thus for fixed e, as n tends to infinity,

$$P(|\frac{S_n}{n} - \mu| \geq e) \rightarrow 0.$$

Thus we have proved that for any $e > 0$, the probability that $\frac{S_n}{n}$ differs from by more than e can be made as small as we please, if n is made large enough. More precisely, we have proved the following theorem:

THEOREM (Law of Large Numbers). Let X_1, X_2,... be an independent trials process. Let $S_n = X_1 + \ldots + X_n$. If $\mu = E(X_j)$, for any $e > 0$

$$P(|\frac{S_n}{n} - \mu| \geq e) \to 0$$

as $n \to \infty$.

Note that $\frac{S_n}{n}$ is an average of the individual outcomes and one often calls the law of large numbers the "law of averages." It is a striking fact that we can start with a random experiment about which little can be predicted and by taking averages obtain an experiment in which the outcome can be predicted with a high degree of certainty.

Example 1. Consider n rolls of a die. Let X_j be the outcome of the jth roll. Then $S_n = X_1 + X_2 + \ldots + X_n$ is the sum of the first n rolls. This is an independent trials process with $E(X_j) = 7/2$. Thus, by the law of large numbers, for any $e > 0$

$$P(|\frac{S_n}{n} - 7/2| \geq e) \to 0.$$

as n tends to infinity. An equivalent way to state this is that for any $e > 0$,

$$P(|\frac{S_n}{n} - 7/2| < e) \to 1.$$

Example 2. Let $X_1, X_2, \ldots$ be a Bernoulli trials process with probability .3 for success and .7 for failure. Let $X_j = 1$ if the jth outcome is success and 0 otherwise. Then, $E(X_j) = .3$ and $V(X_j) = (.3)(.7) = .21$. If

$$A_n = \frac{S_n}{n} = \frac{X_1 + X_2 + \ldots + X_n}{n}$$

then $E(A_n) = .3$ and $V(A_n) = V(S_n)/n^2 = .21/n$. Chebyshev's inequality states that if, for example, $e = .1$,

$$P(|A_n - .3| \geq .1) \leq \frac{.21}{n(.1)^2} .$$

Thus if $n = 100$,

$$P(|A_{100} - .3| \geq .1) \leq .21$$

or if $n = 1000$

$$P(|A_{1000} - .3| \geq .1) \leq .021.$$

These can be rewritten as

$$P(.2 < A_{100} < .4) > .79$$

$$P(.2 < A_{1000} < .4) > .979.$$

The theorem states that as we increase n the probability will approach 1. The same will be true for any choice of e. For example, it will be true for the smaller value $e = .001$. It should be emphasized that while this estimate proves the law of large numbers, it is actually a very crude estimate for the probabilities involved. However, its strength lies in the fact that it is true for any random variable at all and it allows us to prove a very powerful theorem. We can compare the estimates given by Chebyshev's inequality with the exact values by using our program BINOM. We choose instead to write a program LAW which will handle larger values of n. The program LAW computes the exact probability

$$P(.2 < A_n < .4)$$

for n = 50 to 400 in steps of 50.

LAW

```
100 READ P,E
110 FOR N = 50 TO 400 STEP 50
112    LET T = 0
115    FOR J = 0 TO N
130       LET S = 0
140       FOR I = 1 TO J
150          LET S = S+LOG(N-I+1)-LOG(I)
160       NEXT I
170       LET S = S + J*LOG(P) + (N-J)*LOG(1-P)
180       IF J > N*(E+P) THEN 210
190       IF J < N*(E-P) THEN 210
200       LET T = T+EXP(S)
210    NEXT J
220    PRINT N,T
230 NEXT N
240 DATA .3,.1
250 END
```

LAW

```
50              .952236
100             .987501
150             .996509
200             .998991
250             .999709
300             .999915
350             .999974
400             .999992
```

From the run of the program we note, for example, that

$$P(.2 < A_{100} < .4) = .988$$

as compared to the conservative estimate of .79 given by Chebyshev's inequality.

The program LAW uses logarithms to avoid excessively small numbers in computing $B(n,j;.3)$. More specifically we use

$$\log(C(n,j)) = \log\left(\frac{n(n-1)\ldots(n-j+1)}{1\cdot 2\cdot \ldots \cdot j} \right)$$

$$= \sum_{i=1}^{j} \log(n-i+1) - \log(i).$$

$$\log(B(n,j)) = \log(C(n,j)) + j\log(p) + (n-j)\log(1-p).$$

EXERCISES

1. A fair coin is tossed 100 times. Use Chebyshev's inequality to estimate the probability that the number of heads which turns up deviates from the expected number by as much as 3 standard deviations.

   Ans. 1/9.

2. Consider the dice game described at the beginning of Sec. 2. In this game the expected winning per play was found to be -.5. State carefully what the law of large numbers says about a player's average winning in a large number of plays of the game.

3. A fair coin is tossed 500 times. Using Chebyshev's inequality, would you estimate that it is highly likely that the number of heads will not deviate by more than 10 percent from the expected number?

4. Use the program LAW to compute the exact probability that was estimated in Exercise 1. Compare your result with the estimate obtained in Exercise 1.

5. Let S_n be the number of successes in n Bernoulli trials

with probability p for success on each trial. Show, using Chebyshev's inequality, that for any $e > 0$

$$P\left(\left|\frac{S_n}{n} - p\right| \geq e\right) \leq \frac{p(1-p)}{ne^2}.$$

6. For any value of p between 0 and 1, show that

$p(1-p) \leq 1/4$. Using this result, show that the estimate

$$P\left(\left|\frac{S_n}{n} - p\right| \geq e\right) \leq \frac{1}{4ne^2}$$

is valid independent of p.

7. In an opinion poll it is assumed that an unknown proportion p of the people are in favor of a proposed new law and a proportion 1-p are against it. A sample of n people is taken to obtain their opinion. The proportion $\bar{p}$ that answer yes is taken as an estimate of p. Using Exercise 6, how large a sample must be taken if it is desired that the estimate will, with probability .9, be correct to within .01?

8. In Exercise 10 of Section 3, Chap. 2, the hat check problem, you were asked to show that if S_n is the number of people that get their own hats, $E(S_n) = V(S_n) = 1$. Using Chebyshev's inequality show that $P(S_n > 11) \leq .01$ for any n.

9. Let X be any random variable which takes on values $0,1,2,\ldots,n$ and has $E(X) = V(X) = 1$. Show that for any integer value of k

$$P(X \geq k+1) \leq \frac{1}{k^2} .$$

10. Let X be a random variable which takes on the value e with probability 1/2 and -e with probability 1/2. Show that

$$P(|X - E(X)| \geq e) = \frac{V(X)}{e^2} .$$

This shows that given any value of e, we can find a random variable X such that the estimate made by Chebyshev's inequality makes no error. This shows that this is the best possible estimate if we want an inequality that holds for all random variables and all values of e.

11. Let X_1, X_2,.... be any sequence of independent, random variables with expected value 0, but not necessarily having the same distribution. Let

$$S_n = X_1 + X_2 + \ldots + X_n.$$

Assume that $V(S_n) \to 0$. Show that for any $e > 0$

$$P(|S_n| > e) \to 0.$$

12. This problem explores some of the problems of generating random numbers and their relation to checking theorems, like the law of large numbers, by simulation. A common way to generate k digit random numbers between 0 and 1 is the following. Choose three numbers R, A, B. Start with R. Multiply this by A and add B. Remove the integer part and

choose the first k digits of the remaining fractional part of the decimal expansion. This is your first random number. To obtain the second random number repeat this process with R replaced by your first random number RND. To obtain the third random number carry out this procedure with R as your second random number. The procedure is iterated for as many times as you want random numbers. For example, let R = .15, A = 9, and B = 3.2. Then we compute 9(.15) + 3.2 = 4.55. Dropping the 4 our first random number is .55. Then we compute 9(.55) + 3.2 = 8.15 and our second random number is .15. To obtain the third number 9(.15) + 3.2 = 4.55 so that our third number is .55; etc. Write a program to compute random numbers, by the method indicated above, for given values of R, A, B, k. Generate 100 random numbers for k = 3 using several different values of R, A, B. See if you can discover any peculiar properties of your random numbers. Could you reasonably illustrate the law of large numbers by simulating coin tossing?

13. We toss a coin using a random number generator. After each toss we compute the proportion $\frac{S_n}{n}$, where S_n is the number of heads which turn up. Show that the limit, as n tends to infinity, of this ratio is 1.

14. Assume again as in Exercise 13 that we toss a coin using a random number generator. Let S_n now be the number of tails that turn up. Find the limiting probability for the

proportion of tails that will turn up as n tends to infinity.

2. ADDITION OF INDEPENDENT RANDOM VARIABLES

In this section we begin a study of the effect of adding the outcomes of a number of independent chance experiments. This study leads to an explanation of the observed fact that in many natural phenomena, when data are plotted and properly scaled, their graph approximates a bell shaped curve, called the normal curve, defined by the function

$$f(x) = \frac{1}{\sqrt{2\pi}} e^{-x^2/2} .$$

The graph of this function is shown in Figure 1.

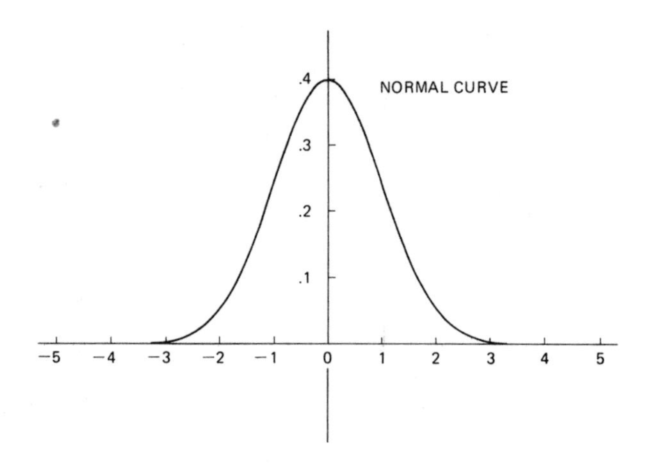

Figure 1.

We begin with a simple experiment which produces a curve

that resembles the normal curve. This experiment is called a Galton desk. A Galton desk consists of a triangular array of nails below which is a row of boxes. A number of balls are started at the top and each ball hits, at each level, a nail and is deflected one unit to the right or left with equal probability. The balls accumulate in the boxes. The particular box in which a ball ends up depends upon the number of deflections made to the right and the number made to the left.

We simulate this experiment for the case of 10 rows of nails and 5000 balls being dropped. A horizontal line is drawn after every fifty balls that fall in the box. We show the process in three stages, first after 1/3 of the balls have dropped, then after 2/3 and finally after all the balls have dropped.

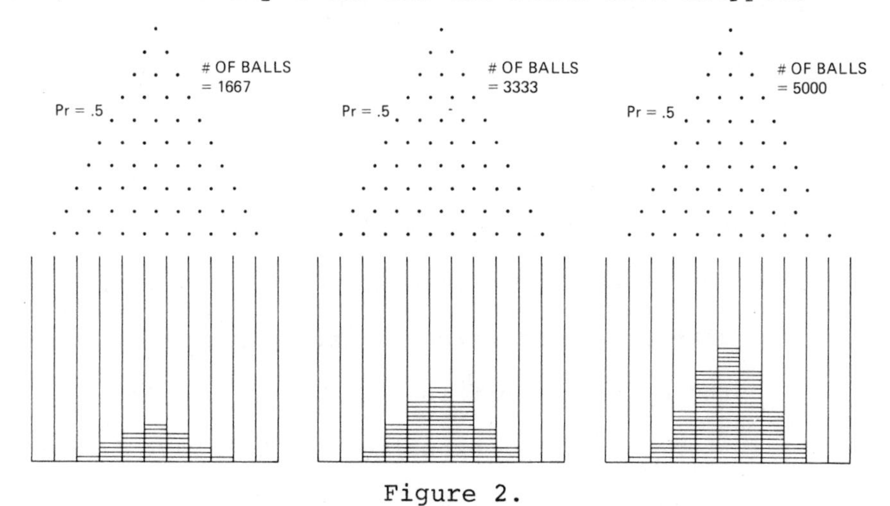

Figure 2.

We note that most of the balls end up in the center box indicating the same number of deflections to the right as to the

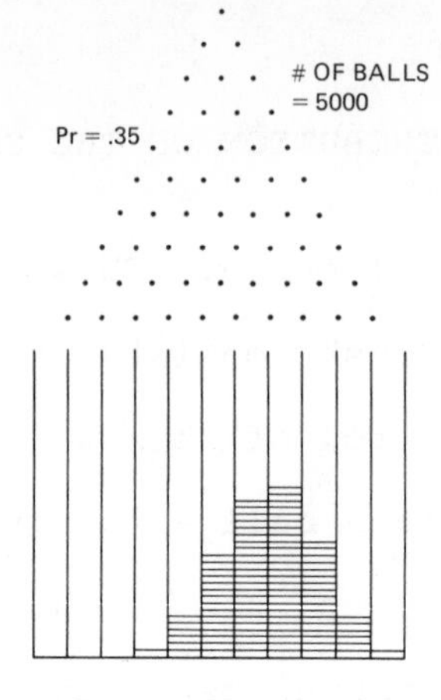

Figure 3.

left. The curve is quite symmetric and in some respects resembles the normal curve.

We next modify our experiment to have a different probability for deflection to the left as compared with that to the right. We have in fact made a deflection each time to the left with probability .35 and to the right with probability .65. In this case we show in Figure 3 the final result only.

We note first that the largest number of balls have fallen into box 2 corresponding to the fact that there is a larger tendency to go to the right. However, even in this unsymmetric case we see that the resulting distribution of balls is rather symmetric and again bears some resemblence to the normal curve.

In each of the experiments we have performed, the final outcome of a single experiment is the result of the sum of a number, namely ten, of small experiments which are independent and take on the value 1 or -1 with probability 1/2,1/2 in the first case and .35,.65 in the second.

We turn then to the study of the distribution of the sum of independent random variables. We shall restrict ourselves to random variables with integer values, positive or negative. The distribution of such a random variable is a set of probabilities $p = \{p_j\}$ defined on the integers. We can assume that p_j is defined for every integer by taking the value to be 0 if it is not a possible value. The following theorem shows how to obtain the distribution of the sum of two independent random variables.

THEOREM. Let X and Y be independent random variables with distribution $p = \{p_j\}$ and $q = \{q_j\}$ respectively. Then the sum $Z = X + Y$ has a distribution $r = \{r_j\}$ with r_j given by

$$(1) \quad r_j = \sum_k p_k q_{j-k}.$$

Proof. The distribution r of Z is given by

$$r_j = P(Z = j) = \sum_k P(X = k, Y = j-k)$$

and since X and Y are independent this is

$$= \sum_k P(X = k)P(Y = j-k) = \sum_k p_k q_{j-k} = r_j.$$

The equation (1) has a simple probabalistic interpretation. In order for the sum Z to have the value j the first outcome X must have some value k and then Y must have the value j-k. We are simply considering all possible values for X and adding the results.

Example 1. A coin is tossed twice. Let X be the number of heads which turn up on the first toss and Y be the number on the second. Let $Z = X + Y$. Then X and Y have the same distribution p. This distribution assigns probability 1/2 to each of the outcomes 0 and 1. Thus by our theorem the distribution for $Z = X + Y$ is given by

$$r_0 = p_0 p_0 = 1/4$$
$$r_1 = p_0 p_1 + p_1 p_0 = 1/2$$
$$r_2 = p_1 p_1 = 1/4.$$

We note that the distribution of Z is the binomial distribution B(2,J). We could have predicted this since it is a special case of Bernoulli trials.

In the last example, we could have avoided the calculation since we know the distribution for the sum of Bernoulli trials. We shall be interested in the more general case of independent trials. Let $X_1, X_2, \ldots$ be such a process. We assume only that X_j takes on integer values, positive or negative. Let

$$S_n = X_1 + X_2 + \ldots + X_n.$$

Let $p^{(n)}$ be the distribution of S_n. That is

$$p_j^{(n)} = P(S_n = j),\ j = 0, \pm 1, \pm 2, \ldots$$

In particular $p^{(1)}$ is the common distribution of X_j and we shall write this more briefly as p. Let $p_j{}^{(n)} = P(S_n = j)$. Then

$$S_n = S_{n-1} + X_n.$$

Since S_{n-1} and X_n are independent, we know from our theorem that $p^{(n)}$ can be obtained from p and $p^{(n-1)}$ by

$$p_j^{(n)} = \sum_k p_k p_{j-k}^{(n-1)}$$

Thus if we know the values of $p^{(n-1)}$ we can obtain the values of $p^{(n)}$. The program SUM carries out this calculation for a given set of p_j's and a given n.

```
SUM

120 DIM Q(100),P(100),R(100)
130 READ R,N
140 FOR I = 0 TO R
150    READ P(I)
160    LET Q(I)=P(I)
170 NEXT I
180 FOR K=2 TO N
190    FOR J=0 TO R*K
200       LET X=0
210       FOR I=0 TO J
220          IF I>R THEN 240
230          LET X = X+P(I)*Q(J-I)
240       NEXT I
250       LET R(J)=X
260    NEXT J
270    FOR I=0 TO R*K
280       LET Q(I)=R(I)
290    NEXT I
300 NEXT K
310 FOR I = 0 TO N*R
320    IF Q(I) = 0 THEN 350
330    PRINT I;
340    PRINT TAB(150*Q(I));"*"
350 NEXT I
360 DATA 6
370 DATA 2
380 DATA 0,.16667,.16667,.16667
390 DATA .16667,.16667,.16667
400 END
```

In this program R is the maximum value for an individual experiment, P(0),P(1),...,P(R) is the basic distribution, and N is the number of experiments. We have made one additional assumption to accommodate the computer. We have assumed that X_j can take on only non-negative integers. As always we assume there are only a finite number of possible outcomes. Let r be the maximum possible outcome on any one experiment. Then to determine the process we need only know the probabilities $p_0, p_1, \ldots, p_r$ which determine the common distribution. Note that we are simply saying that all possible values lie between 0 and r. Not all of these possibilities need occur and so even some of these probabilities may be 0. We put the distribution p in the program as DATA starting in line 380. The values in the program correspond to the experiment of rolling a die. The maximum value r = 6 is put in as a DATA statement in line 360 and the number of times the experiment is to be repeated is put in the DATA statement in line 370. The current value is 2. The Q(I) in line 280 represents $p_i^{(n)}$.

Example 2. A die is rolled n times. We wish to find the distribution for the sum of the numbers which turn up. We consider first the case of just two rolls. In this case we can calculate the probabilities directly by enumerating the 36 possibilities. Each of the possible pairs has probability 1/36.

1 1	2 1	3 1	4 1	5 1	6 1
1 2	2 2	3 2	4 2	5 2	6 2
1 3	2 3	3 3	4 3	5 3	6 3
1 4	2 4	3 4	4 4	5 4	6 4
1 5	2 5	3 5	4 5	5 5	6 5
1 6	2 6	3 6	4 6	5 6	6 6

From this we see

$$P(S_2 = 2) = 1/36$$
$$P(S_2 = 3) = 2/36$$
$$P(S_2 = 4) = 3/36$$
$$P(S_2 = 5) = 4/36$$
$$P(S_2 = 6) = 5/36$$
$$P(S_2 = 7) = 6/36$$
$$P(S_2 = 8) = 5/36$$
$$P(S_2 = 9) = 4/36$$
$$P(S_2 = 10) = 3/36$$
$$P(S_2 = 11) = 2/36$$
$$P(S_2 = 12) = 1/36$$

Note that the probabilities increase linearly up to the maximum value at 7 and then decrease linearly to the minimum value at 12. The data statements in the program SUM are appropriate for this example. We next show a run of the program SUM which gives us the same distribution in graphical form. We can again see the linear relation.

```
SUM

2  *
3       *
4            *
5                 *
6                      *
7                            *
8                      *
9                 *
10           *
11      *
12 *
```

The corresponding calculation for the case of more than 2 sums would be exceedingly tedious. However, to use our program to compute this distribution we need only change the data statement in line 370. Changing this statement from DATA 2 to DATA 3 will give us the distribution for the sum of three rolls of a die.

```
SUM

3 *
4 *
5  *
6    *
7        *
8            *
9               *
10               *
11               *
12              *
13           *
14       *
15   *
16 *
17 *
18 *
```

We note that we already begin to have a curve which shows a similarity to the normal curve.

The essential fact about adding independent experiments is that it tends to smooth out the result in a very regular way. To illustrate this we consider the sum of ten experiments with the probabilities

$$P(X_j = 0) = .4$$

$$P(X_j = 1) = .2$$

$$P(X_j = 2) = .1$$

$$P(X_j = 3) = .3$$

We shall compute the distribution of the sum of ten independent random variables with this common unsymmetric distribution. To do this using the program SUM we change the DATA statements to

```
360 DATA 3

370 DATA 10

380 DATA .4,.2,.1,.3
```

```
          SUM

          0 *
          1 *
          2 *
          3 *
          4 *
          5   *
          6       *
          7           *
          8              *
          9                 *
          10                  *
          11                  *
          12                 *
          13              *
          14           *
          15        *
          16     *
          17  *
          18 *
          19 *
          20 *
          21 *
          22 *
          23 *
          24 *
          25 *
          26 *
          27 *
          28 *
          29 *
          30 *
```

We see from the run of the program for this example that we again obtain a symmetric curve very much like the normal curve.

We thus see both experimentally and by computations that the sum of independent trials can be expected to have a bell shaped distribution similar to that of the normal curve. We have illustrated here the content of the famous central limit theorem of probability. We shall discuss this theorem in the next section.

EXERCISES

1. A game is played as follows. A die is rolled. If the outcome is an even number, the player receives an amount equal to the number on the die. If it is odd, he loses an amount equal to the number on the die. Find the distribution for the player's fortune if he plays the game twice.
2. Using the distribution found in Exercise 1, compute the mean and variance of the player's winnings. Show how these could be obtained from analyzing a single play.
3. For the experiment of the Galton desk find the probability that a given ball will fall in each of the possible boxes, both for the case of a probability .5 for a deflection to the right and for the case of a probability .65 for such a deflection.
4. Two independent experiments are carried out. The first results in an outcome X taking on the values 0,1, and 2 with equal probabilities. The second outcome Y takes on the value 3 with probability 1/4 and 4 with probability 3/4. Find the distribution of (a) $X + Y$ and (b) $X - Y$.

5. A die is rolled three times with outcomes X_1, X_2, X_3. Let Y_3 be the maximum of the values obtained. Show that

$$P(Y_3 \leq j) = P(X_1 \leq j)^3.$$

Use this to find the distribution of Y_3. Does Y_3 have a bell shaped distribution?

6. Let p_n be the probability in n rolls of a die that the maximum number that turns up is j. Show that

$$p_n = (j/6)^n - ((j-1)/6)^n.$$

7. Write a computer program to find the distribution $\{p_j\}$ in Exercise 6 for given n. Run the program for n = 10, 20, 30 and make a conjecture on the limit of p_j as j --> 0.

8. Write a program to simulate the following Galton desk. Assume that there are 30 deflections. Each is one unit to the right with probability p or one unit to the left with probability q = 1-p. Have the program print out a graph using the tab function which will represent the relative numbers which end up in each box. Run the program for the case of 5000 balls first for p = 1/2 and then for p = .65. Compare the results in these two cases.

9. Modify the program written in Exercise 8, to print out the proportion of balls that fall in each box. Compare this <u>empirical distribution</u> for the final position of an individual ball with the exact distribution.

10. Let X and Y be random variables with integer values which can take on at least one negative value. Let a be the smallest value that X can take on and b the smallest value that Y can take on. Then the random variables $X^* = X - a$ and $Y^* = Y - b$ take on non-negative values. Show that if the distribution of $Z^* = X^* = Y^*$ is known then so is the distribution of $Z = X + Y$.

11. A die is rolled three times. Let S_3 be the total number of points that turn up. Using (1) and Example 2, find $P(S_3 = 10)$.

The following exercises relate to the concept of generating functions. Let X be a random variable which takes on one of a finite number of non-negative integers, 0, 1, 2,..., r, and with distribution $p = \{p_j\}$. The <u>generating function</u> of X is the polynomial defined by

$$f(x) = \sum_{j=0}^{r} p_j x^j.$$

12. Find the generating function for the random variable which gives the number of successes on the jth outcome of Bernoulli trials with probability p for success and q for failure.

Ans. $f(x) = p + qx$

13. Let X and Y be independent random variables with generating functions $f(x)$ and $g(x)$ respectively. Show that the generating function of the sum $Z = X + Y$ is the function $h(x) = f(x)g(x)$.

14. Using the results of Exercises 12 and 13, find the generating function for the number of successes in n Bernoulli trials.

15. Let $f(x)$ be the generating function of a random variable X. Show that (a) $f(1) = 1$. (b) $E(X) = f'(1)$. (c) $E(X^2) = f''(1) - f'(1)$. (d) $V(X) = f''(1) - f'(1) - (f'(1))^2$. ($f'(x)$ and $f''(x)$ are the first and second derivatives of $f(x)$).

3. THE CENTRAL LIMIT THEOREM

We have seen that the sum, $S_n = X_1 + X_2 + \ldots + X_n$ of independent random variables with a common distribution has a distribution whose graph approximates a bell-shaped curve. In this section we shall give a precise meaning to this. We first consider some examples. In graphing the distributions in this section we shall use the program SUM to compute the distribution, but we have put the output into a graphical display device in order to give better pictures for the distribution.

<u>Example 1.</u> Let $S_n = X_1 + X_2 + \ldots + X_n$ where X_j takes on the value 0 or 1. Let $p = P(X_j = 1)$. Then we know from the study of independent trials that S_n has expected value np and variance npq. Also the distribution of S_n is given by

$$p_j^{(n)} = P(S_n = j) = \binom{n}{j} p^j q^{n-j}.$$

In Figure 4 we have graphed this distribution for $p = .5$ and $n = 10,50,100$.

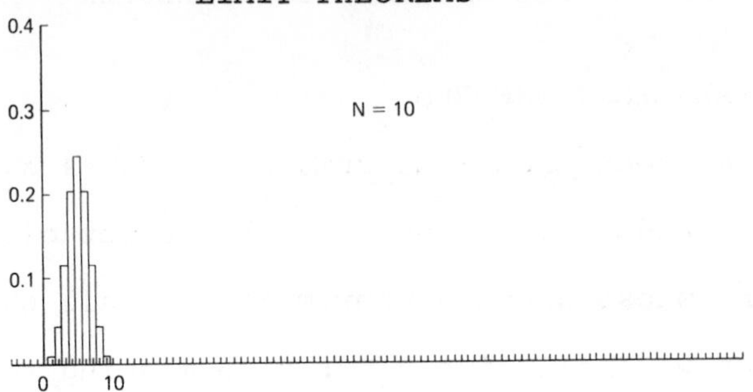

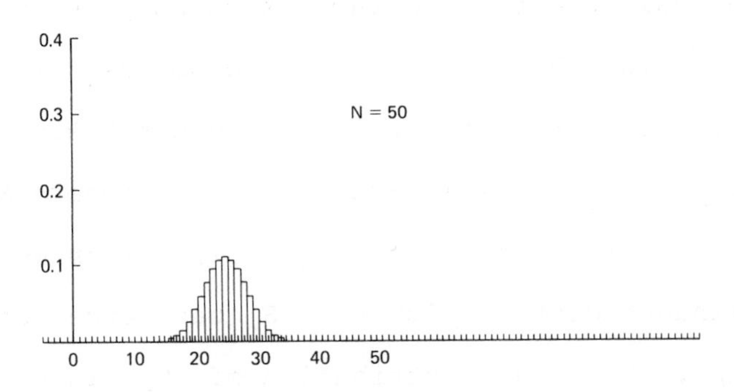

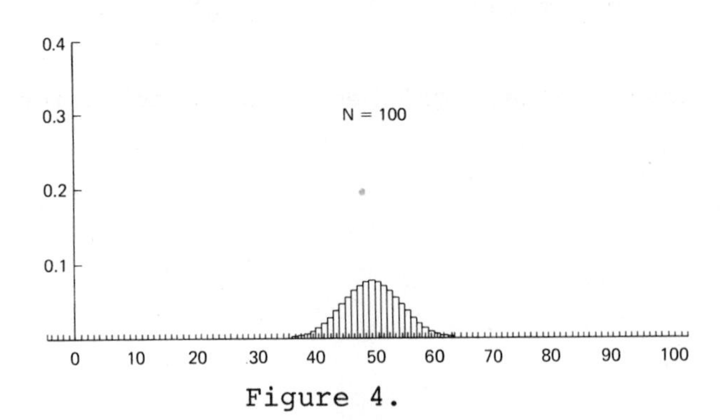

Figure 4.

We note that the peak of the curve drifts to the right and the spread increases.

Our goal is to compare the distribution of a general sum, S_n, of independent random variables with one fixed bell-shaped curve, the normal curve, introduced in the previous section. We have just seen that we cannot hope to do this unless we at least scale the distribution to prevent it from flattening and drifting away. In our examples from the last section we can see that the most likely values for the sum of independent random variables occur near the expected value. The law of large numbers would, of course, suggest this. Thus it is natural to center the distribution in terms of the expected value of the sum. If $E(X_1) = \mu$ then $E(S_n) = n\mu$. We avoid the drift by changing S_n to $S_n - n\mu$. This new random variable will have expected value 0. Next we know that the variance is a measure of the spread of a distribution. To prevent the distribution from spreading we are going to normalize S_n to always have a fixed variance of 1. If $V(X_1) = \sigma^2$ then $V(S_n) = n\sigma^2$ and $D(S_n) = (n\sigma^2)^{1/2}$. To make a random variable have variance 1 we need only divide by its standard deviation. Thus we change $S_n - n\mu$ to

$$S_n^* = \frac{S_n - n\mu}{\sigma\sqrt{n}} .$$

We have thus produced from S_n a <u>standardized random variable</u>, S_n^*, which has expected value 0 and variance 1. This holds for all n.

If x is a possible outcome of S_n, the corresponding outcome for S_n^* will be

$$x^* = \frac{x - n\mu}{\sigma\sqrt{n}} .$$

Example 1. A fair coin is tossed 16 times. Let S_{16} be the number of heads which turn up. Then $E(S_{16}) = 8$ and $V(S_n) = 16(1/2)(1/2) = 4$. Thus $D(S_n) = 2$ and

$$S_{16}^* = \frac{S_{16} - 8}{2}.$$

Thus the possible values and their probabilities of occurring are given in the following table.

x	x*	$P(S_{16} = x) = P(S_{16}^* = x^*)$
0	-4	.0000
1	-3.5	.0002
2	-3	.0018
3	-2.5	.0085
4	-2	.0278
5	-1.5	.0667
6	-1	.1222
7	- .5	.1746
8	0	.1964
9	0.5	.1746
10	1	.1222
11	1.5	.0667
12	2	.0278
13	2.5	.0085
14	3	.0018
15	3.5	.0002
16	4	.0000

We note that the distribution of S_{16}^* is centered at 0 and has a spread from -4 to 4. We note also that the individual probabilities for the outcomes of S_{16} are relatively small. For larger n they would become very small. For this reason we shall be interested not so much in the probability of a specific outcome, but rather in the probability that the outcome lies in an interval. That is, in probabilities of the form

$$P(a < S_n^* < b)$$

The central limit theorem will relate these probabilities to the area under the normal curve between a and b. Before discussing this theorem, we shall first review the definition of the area under a curve between two limits. We shall do this in terms of finding the area under the normal curve between -2 and 2. The procedure for an arbitrary interval is the same. We begin by choosing $n + 1$ points $x_0, x_1, x_2, \ldots, x_n$ in the interval $(-2,2)$. We choose $x_0 = -2$, $x_n = 2$ and make the remaining points an equal distance apart. The length of each interval is then $d = 4/n$. At each point x_j we construct a rectangle with base of length d centered at x_j and height equal to $f(x_j)$. The area of the rectangle centered at x_j is then $d \cdot f(x_j)$. The sum of the areas of these rectangles

$$N_n(-2,2) = \sum_j d \cdot f(x_j)$$

is an approximation to the area, $N(-2,2)$, under the normal curve between -2 and 2. We illustrate this in Figure 5.

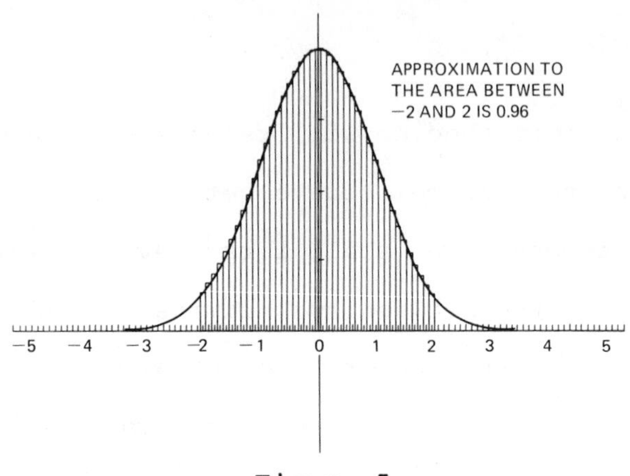

Figure 5.

As we increase n indefinitely, the numbers $N_n(-2,2)$ tend to a limit, $N(-2,2)$, called the <u>area</u> under the curve between -2 and 2. The procedure for defining the area under the curve between a and b for any a and b with $a < b$ is the same. We shall denote the area under the normal curve between a and b by $N(a,b)$. Those that have had calculus will recall that the area under a curve $f(x)$ between a and b is given by

$$A(a,b) = \int_a^b f(x)dx$$

In calculus one learns to find the area under most simple curves without using the approximation technique used in the definition. However, the normal function is one of the few simple functions for which numerical approximations are required. We could use the method indicated above, but the following program presented in Kemeny and Kurtz "Basic Programming" is more efficient.

```
AREA

100 READ A,B
110 DEF FNF(X) = (1/SQR(2*3.141592*EXP(-(X↑2)/2)
120 LET H = B-A
130 LET T = (FNF(A)+FNF(B))*H/2
150 GO TO 170
160 LET T = (T+M)/2
170 LET M = 0
180 FOR X = A +H/2  TO B STEP H
190    LET M = M+FNF(X)
200 NEXT X
210 LET M = M*H
220 LET S = (T+2*M)/3
230 LET H = H/2
240 IF ABS(T-M)/ABS(S) > .00001 THEN 160
250 PRINT "AREA UNDER THE NORMAL CURVE BETWEEN ";
255 A;"AND";B;"EQUALS";S
260 DATA -2,2
270 END
```

```
AREA

AREA UNDER THE NORMAL CURVE BETWEEN -2 AND 2 EQUALS 0.9545
```

While it is easy to use a program like AREA for a specific area, we shall use also the following table to find areas under the normal curve.

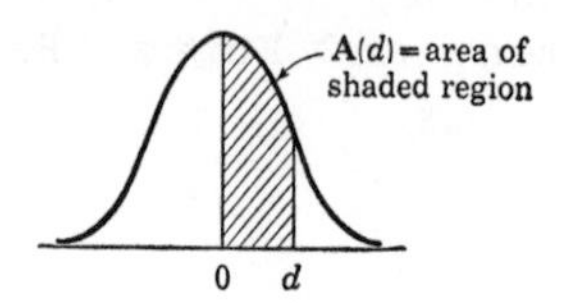

d	A(d)	d	A(d)	d	A(d)	d	A(d)
.0	.000	1.1	.364	2.1	.482	3.1	.4990
.1	.040	1.2	.385	2.2	.486	3.2	.4993
.2	.079	1.3	.403	2.3	.489	3.3	.4995
.3	.118	1.4	.419	2.4	.492	3.4	.4997
.4	.155	1.5	.433	2.5	.494	3.5	.4998
.5	.191	1.6	.445	2.6	.495	3.6	.4998
.6	.226	1.7	.455	2.7	.497	3.7	.4999
.7	.258	1.8	.464	2.8	.497	3.8	.49993
.8	.288	1.9	.471	2.9	.498	3.9	.49995
.9	.316	2.0	.477	3.0	.4987	4.0	.49997
1	.341					5.0	.499999977

It is clear from the symmetry of the normal curve that areas such as that between -2 and 3 can be found from this table using the fact that the area from -2 to 0 is the same as that from 0 to 2.

We turn now to the way in which the normal curve relates to the distribution of the sum of independent trials. We shall show this first for the case of Bernoulli trials. For this case, if S_n is the number of successes in n experiments, then the standardized sum is

$$S_n^* = \frac{S_n - np}{\sqrt{npq}} .$$

We want to estimate a probability of the form

$$\sum_{a \leq x^* \leq b} P(S_n = x^*).$$

The values of x^* which lie between a and b are at a distance $d = 1/\sqrt{npq}$ apart. At each point x^* we construct a rectangle centered at x^*, and with area equal to $P(S_n = x^*)$. We do this by making the length of the base d and the height

$$P(S_n = x^*)/d.$$

The central limit theorem then states that the sum of these rectangles also approximate the area N(a,b) under the normal curve between a and b.

THEOREM. (Central Limit Theorem for Bernoulli trials). Let S_n be the number of successes in n Bernoulli trials with probability p for success on each trial. Then

$$P\left(a < \frac{S_n - np}{\sqrt{npq}} < b\right) \to N(a,b).$$

Example 1. (Continued). In Figure 6 we have plotted the standardized distributions for the Bernoulli trials distributions plotted in Figure 4. We have superimposed the normal curve. We see that our normalization has prevented the drift and spread in the distributions. We see also that the approximation to the normal curve is quite good.

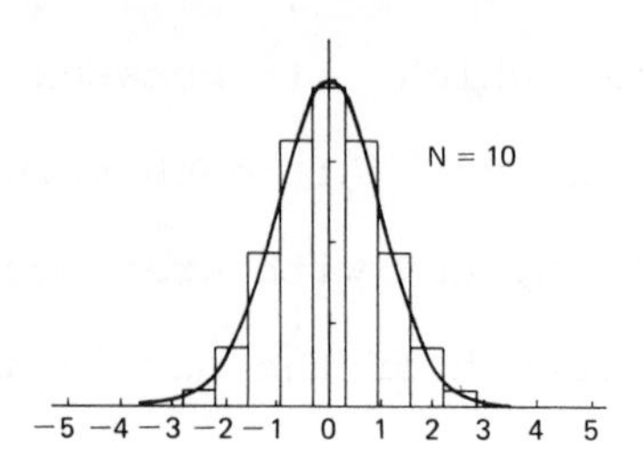

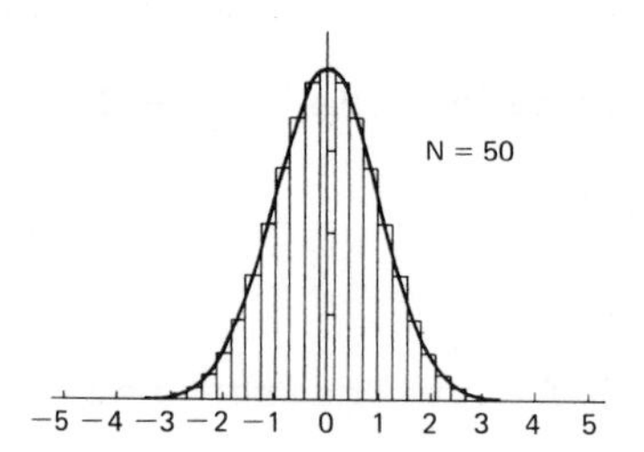

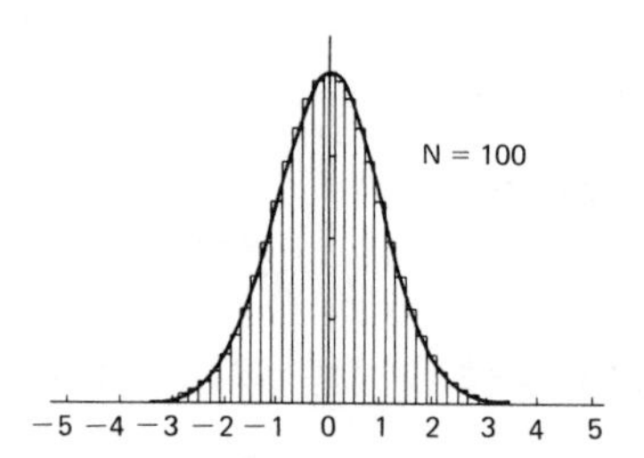

Figure 6.

The distributions in Figure 6 correspond to the symmetric Bernoulli distribution with $p = .5$. In Figure 7 we give a corresponding comparison with the normal curve for a standardized distribution starting with the unsymmetric distribution with $p = .2$ and $n = 30$.

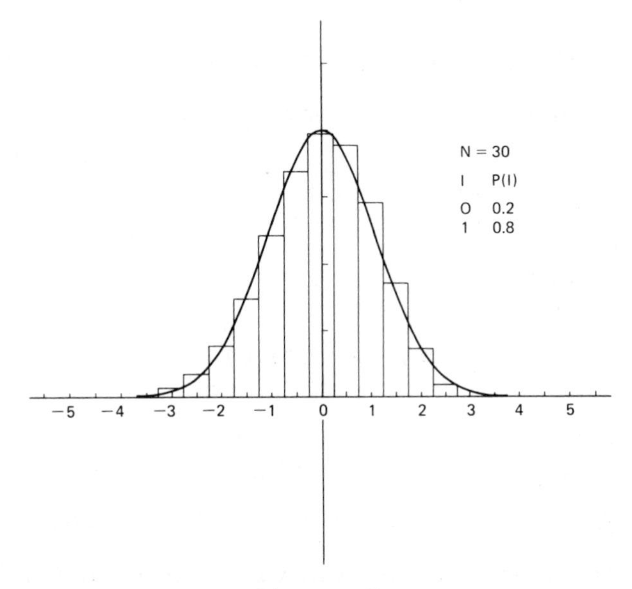

Figure 7.

It is quite remarkable that a symmetric curve like the normal curve can fit such an unsymmetrical distribution so well. Even more remarkable is the fact that the central limit theorem applies in much greater generality. While still far from the general result, the central limit theorem holds equally well for general independent trials processes in which each outcome has only a finite number of possible values.

THEOREM. Let $X_1, X_2, \ldots, X_n$ be an independent trials process with $\mu = E(X_j)$ and $\sigma^2 = V(X_j)$. Let $S_n = X_1 + \ldots + X_n$. Let

$$S_n^* = \frac{S_n - \mu}{\sigma\sqrt{n}} .$$

Then for any a and b

$$P(a < S_n^* < b) \to N(a,b)$$

as n tends to infinity.

<u>Example 2</u>. Let us return to the experiment that we considered in the previous section with outcomes 0,1,2,3 occuring with probabilities .4,.2,.1,.3. We made a crude graph of the distribution of the sum of 30 experiments with this common distribution. In Figure 8 we have graphed the rectangles whose areas represent the probabilities for $S_{16}^* = x^*$. We see that even for $n = 16$, a relatively small value of n, the distribution closely fits the normal curve as suggested by the central limit theorem.

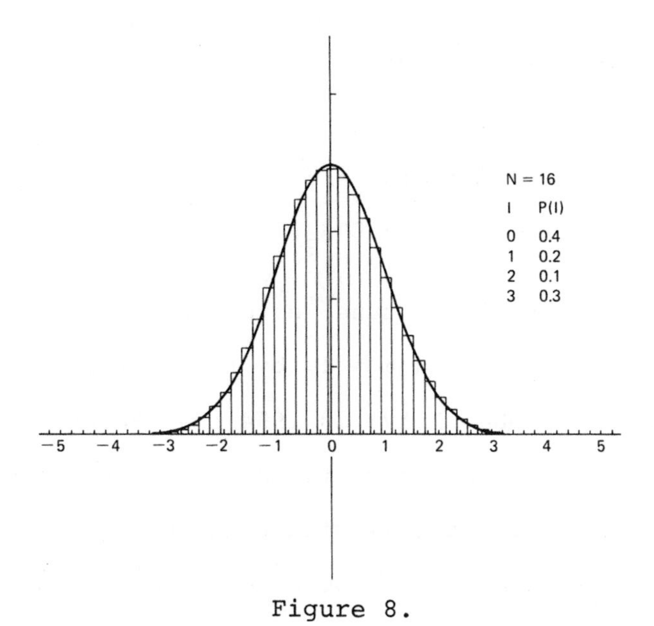

Figure 8.

In using the central limit theorem, it is usual to think in units of standard deviations. In fact, we can rewrite (3) as

$$P(a\sigma\sqrt{n} \leq S_n - \mu \leq b\sigma\sqrt{n}) \rightarrow N(a,b).$$

Thus, for example, the probability of a deviation of no more than k standard deviations from the expected value is approximated by the area under the normal curve from -k to k. From our table of areas we see that for k = 3 this probability is approximately .997. For k = 2 it is .954. For k = 1 it is .681. Thus a deviation of 3 or more standard deviations from the mean is very unlikely. However, a deviation of as much as one standard deviation would not be surprising.

Example 3. Consider our game of matching pennies. Let X_j be Peter's winning on the jth toss. Then

$$P(X_j = 1) = 1/2$$

$$P(X_j = -1) = 1/2.$$

$$E(X_j) = (1)(1/2) + (-1)(1/2) = 0$$

$$V(X_j) = E(X_j^2) - E(X_j)^2 = 1$$

Thus, if S_n is Peter's fortune after n matches, $E(S_n) = 0$, $V(S_n) = n$. Since $\mu = 0$ and $\sigma = 1$, by the central limit theorem, in 900 matches the probability is .954 that Peter's fortune will not be more than 2 standard deviations, i.e. $2\sqrt{900} = 60$ away from the mean 0. That is, it will with high probability be between -60 and 60. This example is also a one-dimensional random walk or as it is sometimes called, a drunkard's walk. The drunkard proceeds by choosing at each corner to go forward one block with probability 1/2 or back one block with probability 1/2. Thus if he starts from home, after 900 decisions we do not expect to find him more than 60 blocks from home.

Of course he may have a bias to go one way, say to the right, if there is another bar 100 blocks down the road. Assume in fact that he moves to the right with probability .6 and to the left with probability .4. Then $E(X_j) = .6 - .4 = .2$ and $V(X_j) = E(X_j^2) - E(X_j)^2 = 1 - .04 = .96$. Thus $E(S_{900}) = 180$ and $V(S_{900}) = (900)(.96) = 864$ and $\sigma\sqrt{n} = \sqrt{(900)(.96)} = 30\sqrt{.96} = 29.3$. Thus we would expect after 900 decisions to find him not more than two standard deviations or 59 blocks from the expected position 180 blocks to the right of his starting point. Thus if our model is correct he would have walked by the bar.

<u>Example 4.</u> Assume that an Ivy League School finds that about 1/2 of the students admitted accept. The college can accommodate no more than 830 students. If it sends out 1600 acceptances, the expected number of acceptances is 800. The standard deviation is $\sqrt{npq} = \sqrt{1600(1/2)(1/2)} = 20$. We are now only interested in finding the probability that too many students accept, in particular,

$$P(S_{1600} \geq 830).$$

But 830 is 1.5 standard deviations more than the mean. Thus we have

$$P(S^*_{1600} \geq 1.5) \doteq .067.$$

Thus the procedure is relatively safe if the independent trials model is correct. The central limit theorem shows that we can make relatively accurate predictions about sums of identically distributed random variables. If we want the sum of our random variables to be close to some target figure, we know by the central limit theorem that if we have enough random variables, their sum will be relatively close to the target. It is important to realize , however, that the theorem also gives us limits of precision. For example, a deviation of less than .1 of a standard deviation from the mean is unlikely, that is $P(S^*_n <$ $.1) = .040$. Thus if an admissions officer admits 1600 students, and is regularly within 1 or 2 of the goal of 800, we can assume that the independent trials model is not appropriate for acceptances.

EXERCISES

1. Let S_{100} be the number of heads which turn up in 100 tosses of a coin. Use the central limit theorem to estimate
   (a) $P(S_{100} < 45) =$
   (b) $P(44 < S_{100} < 56) =$
   (c) $P(S_{100} > 63) =$
2. A rookie, Mickey Mantle, is brought to a ball club on the assumption that he will have a .300 batting average. In his first year he comes to bat 300 times and his batting average is .267. If we assume that he has probability .3 of getting a hit every time, is it plausible that he could do this badly, or should he be sent back to the minor leagues?
3. Once upon a time there were two railway trains competing for the passenger traffic of 1000 people leaving from Chicago at the same hour going to Los Angeles. Assume that passengers are equally likely to choose each train. How many seats must a train have to have probability .99 or better of having a seat for a passenger?
4. A piece of rope is made up of 100 strands. Assume that the breaking strength of the rope is the sum of the breaking strength of the individual strands. Assume further that this sum may be considered to be the sum of an independent trials

process with 100 experiments each having expected value of 10 pounds and standard deviation 1. Find the approximate probability that the rope will support a weight (a) of 1000 pounds, (b) of 970 pounds.

Ans. (a) .5; (b) .999.

5. Write a program to find the sum of 10,000 random integers 0,1,2,3,4,5,6,7,8, or 9. Compute the average of the numbers obtained. Have the program test to see if the average lies within 3 standard deviations of the expected value of 4.5. Modify the program so that it repeats 1000 times and keeps track of the number of times the test is passed.

6. A man goes to work each morning by bus. His bus is 5 minutes late 10 percent of the time, 10 minutes late 40 percent of the time, and 15 minutes late 50 percent of the time. If he takes the bus 230 mornings a year, is it likely that he has to wait as much as 50 hours?

7. A space trip is planned which will take 400 hours. An essential part of the spaceship has a lifetime distribution, measured in hours, with mean 10 and standard deviation 2. How many parts should be taken if it is desired to have a probability of at least .999 of being able to replace this part each time it needs replacement?

8. Consider a Bernoulli trials process with outcomes A and B. Let p be the probability for outcome A and q = 1-p the probability for B. Assume that p is not known and that it

is desired to estimate p to be the average number of times A occurs in n trials. Using the fact that $\sqrt{pq} \leq .5$ estimate the number of trials necessary to ensure that the probability is at least .95 that the fraction of outcomes which are A will be within .02 of the unknown value for p.

9. Let S be the number of heads in 1,000,000 tosses of a balanced coin. Use (a) Chebyshev's inequality, and (b) the central limit theorem to estimate the probability that S is less than 499,500 or greater than 500,500. Use the same two methods to estimate the probability that S is less than 499,000 or greater than 501,000. And also the probability that S is less than 498,500 or greater than 501,500.

   Ans. (a) ≤ 1 (i.e., no information); $\leq 1/4$; $\leq 1/9$.
   (b) .318; .046; .003 (approximate answers).

10. A true-false examination has 48 questions. Assume that the probability that a student knows the answer to any one question is 3/4. A passing score is 30 or better. Estimate the probability that a student will fail the exam.

The following questions relate to a method of estimation used in statistics to estimate an unknown parameter in a process such as a Bernoulli trials process when the probability p for success is not known.

11. Consider a Bernoulli trials process with probability p for success on each trial. Let S_n be the number of sucesses in n trials and $\bar{p} = S_n/n$ be the proportion of successes. Show that the central limit theorem leads to the approximation

$$P(-2\sqrt{pq/n} \leq \bar{p} - p \leq 2\sqrt{pq/n}\ ) \doteq .95$$

Using the fact that $\sqrt{pq} < 1/2$ show that the probability is $\geq$ .95 that the inequality

$$\bar{p} - \frac{1}{\sqrt{n}} < p < \bar{p} + \frac{1}{\sqrt{n}}$$

will hold. Note that p is fixed and $\bar{p}$ is the random quantity. The interval $(\bar{p} - 1/\sqrt{n} \leq p \leq \bar{p} + 1/\sqrt{n})$ is called a <u>95 percent confidence interval</u> for p.

12. Write a computer program to simulate 10,000 Bernoulli trials with probability .3 for success on each trial. Have the program compute the 95 percent confidence interval for the probability of success based only on the proportion of successes. Repeat the experiment 20 times and see how many times the true value of .3 is included within the confidence limits.

13. A die has the number 1 on two opposite faces, the number 2

on two opposite faces, and the number 3 on two opposite faces. The die is rolled 216 times. Find the probability that the sum of the outcomes is greater than 456.

4. THE POISSON APPROXIMATION

In this section we consider a second limit theorem less general than the central limit theorem, but of interest in a number of applications.

Let S_n be the number of successes in n Bernoulli trials with probability p for success on each trial. Then $E(S_n) = np$. To obtain the central limit theorem we changed the units to obtain a new random variable S_n^* which had, for all n, the same expected value and variance. We shall now keep the same units but change the value of p to keep the expected value $m = np$ a constant.

THEOREM. (Poisson limit theorem). Let S_n be the number of successes in a Bernoulli trials process with probability p for success on each trial. Assume that n approaches infinity and p approaches 0 in such a way that $np = m$ remains constant. Then

$$P(S_n = j) \to \frac{m^j e^{-m}}{j!} .$$

Proof: Consider the case $j = 0$. For this case

$$(1)\quad P(S_n = 0) = (1-p)^n = \left(1 - \frac{m}{n}\right)^n .$$

It is shown in calculus that for any value a

$$(2)\quad \left(1 + \frac{a}{n}\right)^n \to e^a$$

as n tends to infinity. Choosing $a = -m$ in (2) we obtain

$$P(S_n = 0) \to e^{-m}$$

as n tends to infinity. This proves the theorem for $j = 0$. In general for $j \geq 1$, let $q = 1 - p$. Then

$$\frac{P(S_n = j)}{P(S_n = j-1)} = \frac{\binom{n}{j} p^j q^{n-j}}{\binom{n}{j-1} p^{j-1} q^{n-j+1}}$$

$$= \frac{n-j+1}{j} \frac{p}{q} = \frac{m - jp + p}{j(1-p)} .$$

Since $p \to 0$ and m and j are fixed we have

$$\frac{P(S_n = j)}{P(S_n = j-1)} \to \frac{m}{j} \quad \text{as} \quad n \to \text{infinity}.$$

Thus

$$P(S_n = 1) = \frac{P(S_n = 1)}{P(S_n = 0)} P(S_n = 0)$$

$$\to me^{-m}.$$

Similarly

$$P(S_n = 2) \to \frac{m}{2} me^{-m} = \frac{m^2}{2!} e^{-m}$$

$$P(S_n = 3) \to \frac{m}{3} \frac{m^2}{2!} e^{-m} = \frac{m^3}{3!} e^{-m} ,$$

etc. In general

$$P(S_n = j) \to \frac{m^j}{j!} e^{-m}.$$

Example 1. A typesetter makes, on the average, one mistake per 1000 words. Assume that he is setting a book with 100 words to a page. Let S_{100} be the number of mistakes that he makes on a single page. Then the exact probability would be obtained by considering S_{100} the result of 100 Bernoulli trials with $p = 1/1000$. The expected value of S_{100} is $m = 100(1/1000) = .1$. The Poisson approximation is then

$$P(S_{100} = j) \doteq \frac{e^{-.1}(.1)^j}{j!} .$$

We give below the exact values computed by the Binomial distribution and the Poisson approximation.

	Exact	Poisson
0	.90480	.90484
1	.09057	.09048
2	.00449	.00452
3	.00015	.00015
4	.00000	.00000

Example 2. A number of applications of this approximation are equivalent to the following somewhat frivolous application. Assume that we are making raisin cookies. We put a box of 600 raisins into our dough mix, mix up the dough, then make from the dough 500 cookies. We then ask for the probability that a randomly chosen cookie will have 0,1,2,.. raisins. To compute this probability we assume that each of the 600 raisins has an equal probability of being in a particular cookie. This probability would be $p = 1/500$. Then we may regard the number of raisins in a cookie to be the result of $n = 600$ independent trials with probability $p = 1/500$ for success on each trial.

Since n is large and p is small we can use the Poisson approximation with $m = 600(1/500) = 1.2$. The exact probabilities and the Poisson approximations for the number of raisins in a given cookie are given in Figure 9. We note that the model does not suggest that the raisins are likely to cluster in a few cookies. It is extremely unlikely that any one cookie will have as many as five raisins. Examples similar to this but relating to more serious matters such as blood counts can be found in Feller.

	Exact probability	Poisson approximation
0	.3008	.3012
1	.3617	.3614
2	.2171	.2169
3	.0867	.0867
4	.0259	.0260
5	.0062	.0062
6	.0012	.0012
7	.0002	.0002
8	.0000	.0000

Figure 9.

In using the Poisson approximation we replace the distribution

$$(2) \quad p_j^{(n)} = \binom{n}{j} p^j q^{n-j}, \quad j = 0,1,\ldots,n$$

by the numbers

$$(3) \quad p_j = \frac{m^j e^{-m}}{j!}, \quad j = 0,1,2,\ldots$$

where $m = np$. In Figure 10 we have given three comparisons of these approximations. We see that when p is small they are quite good.

	Poisson $m = .1$	Binomial $n = 10$ $p = .1$	Poisson $m = 1$	Binomial $n = 100$ $p = .01$	Poisson $m = 10$	Binomial $n = 1000$ $p = .01$
0	.9048	.9044	.3679	.3660	.0000	.0000
1	.0905	.0914	.3679	.3697	.0005	.0004
2	.0045	.0042	.1839	.1849	.0023	.0022
3	.0002	.0001	.0613	.0610	.0023	.0022
4	.0000	.0000	.0153	.0149	.0189	.0186
5			.0031	.0029	.0378	.0374
6			.0005	.0005	.0631	.0627
7			.0001	.0001	.0901	.0900
8			.0000	.0000	.1126	.1128
9					.1251	.1256
10					.1251	.1257
11					.1137	.1143
12					.0948	.0952
13					.0729	.0731
14					.0521	.0731
15					.0347	.0345
16					.0217	.0215
17					.0128	.0126
18					.0071	.0069
19					.0037	.0036
20					.0019	.0018
21					.0009	.0009
22					.0004	.0004
23					.0002	.0002
24					.0001	.0001
25					.0000	.0000

Figure 10.

We note that the numbers p_j used in the Poisson approximation are defined for all non-negative j. Further if p_j is given by (3), we have

$$\sum_j p_j = e^{-m} \sum_{j=0} \frac{m^j}{j!} = e^{-m} e^m = 1.$$

Here we have used the fact that

$$e^x = \sum_{j \geq 0} \frac{x^j}{j!} .$$

We can thus interpret the Poisson measure as assigning a set of probabilities to the infinite set 0,1,2,..., consisting of all non-negative numbers. The probability of events, expected values of random variables, and other such concepts can be defined as in the finite case. We shall not go further with this but suggest some of these calculations in the exercises.

EXERCISES

1. Assume that the Poisson measure with mean .3 has been assigned for the outcome of an experiment. Let X be the outcome function. Find $P(X = 0)$, $P(X = 1)$ and $P(X > 1)$.
2. Assume that on the average only one person in a thousand has a particular rare blood type.

   (a) In a city of 10,000, what is the probability that no person has this blood type?

   (b) How many people would have to be canvassed to give a probability greater than one half of finding at least one person with this blood type?
3. The probability that in a bridge deal one of the four hands has all hearts is $(6.3)(10^{-12})$. The probability expert in a

town of 50,000 is called, on the average of once a year (usually late at night), and told that the caller has just been dealt a hand of all hearts. Should he suspect that some of these callers are the victims of practical jokes?

4. A man never puts money in a five-cent parking meter. He assumes that there is a probability of .05 that he will be caught. The first offense costs nothing, the second costs 50 cents, and subsequent offenses cost one dollar each. Under his assumptions, how does the expected cost of parking 20 times compare with the cost of putting money in the meter each time?

5. An advertiser drops 10,000 leaflets on a city which has 2,000 blocks. Assume that each leaflet has an equal chance to land on each block. What is the probability that a particular block will receive no leaflets?

6. Assume that for a certain experiment the Poisson measure with mean m has been assigned. Show that a most probable outcome for the experiment is the value k such that $m-1 \leq k \leq m$. Under what conditions will there be two most probable values?

7. A man receives an average of ten letters each day. On a certain day he receives no mail and wonders if it is a holiday. To decide this he computes the probability that in ten years he would have at least one day without any mail. He assumes that the number of letters he receives on a day

has a Poisson measure. What probability did he find? (Hint: Apply the Poisson measure twice. First, to find the probability that on a given day he receives no mail, and a second time to find the probability that in 3000 days he will have no such day--since each year has about 300 days on which mail is delivered.)

8. In Example 1, assume that the book has 1000 pages. Let X be the number of pages with no mistakes. Show that $E(X) = 905$ and $V(X) = 86$. Using this, show that the probability is $\leq$ .05 that there will be more than 950 pages without errors or fewer than 860 pages without errors.

9. A Poisson measure has been assigned to the non-negative integers by $p_j = \frac{m^j e^{-m}}{j!}$. We let X be the outcome. That is, $P(X = j) = p_j$. We define as in the finite case

$$\mu = E(X) = \sum_j jp_j.$$

and $\sigma^2 = V(X) = \sum_j (j - m)^2 p_j$.

Prove that $\mu = m$ and $\sigma^2 = m$.

10. For an independent trials process let X be the first time a success occurs. Find

$$P(X = s + k | X > s),\ k = 1, 2, \ldots.$$

Interpret your answer.

11. A perpetual crap game goes on at Charley's. On each play a pair of dice is rolled. Find the expected number of plays

between occurrences of snake eyes, i.e., a pair of ones.

12. Jones comes into Charley's on an evening when there have already been 100 plays. He plans to play until the first occurrence of snake eyes. What do you estimate to be the expected time that he will play? Write a computer program to simulate Jones' time of play. By repeating this experiment a large number of times, estimate the average number of plays Jones will make. How does the computer estimate compare with your estimate?

13. Feller discusses the statistics of flying bomb hits in the south of London during the second world war. The area was divided into 24x24 = 576 small areas. The total number of hits was 537. Thre were 229 squares with 0 hits, 211 with 1 hit, 93 with 2 hits, 35 with 3 hits, 7 with 4 hits and 1 with 5 or more. Assuming the hits were purely random, use the Poisson approximation to find the probability that a particular square would have exactly k hits. Compute the expected number of squares that would have 0, 1,2,3,4, and 5 or more hits and compare this with the observed results.

14. Simulate the situation desxribed in Exercise 13. Print out the 24x24 matrix giving the actual number of bombs which would fall in each square with your simulation. Do you observe a tendency for the hits to cluster?

15. An airline find that 4 percent of the passengers that make reservations on a particular flight will not show up.

Consequently, their policy is to sell 100 reserved seats on a plane that has only 98 seats. Find the probability that every person that shows up for the flight will find a seat available.

16. The king's minter boxes his coins 500 to a box, and puts one counterfeit coin in each box. The king is suspicious, but instead of testing all the coins in one box, he tests one coin chosen at random out of each of 500 boxes. What is the probability that he finds at laest one fake? What if the king tests one coin from each of 1000 boxes?

5*. PENNY MATCHING REVISITED

We once more come upon a penny matching game. This time it is between Mary and John. We recall that on each match Mary wins one penny with probability 1/2 and loses one penny with the same probability. Let S_n be Mary's fortune after n matches. Recall that we can also consider S_n to be the position of a random walk which moves on the integers making, on each step, a step to the right with probability 1/2 or to the left with probability 1/2. This interpretation is sometimes called simple random walk. By the law of large numbers, S_n/n approaches 0 as n tends to infinity. By the central limit theorem, with high probability S_n itself will not lie more than $3\sqrt{n}$ units away from 0. But what about Mary's fortune throughout the sequence of plays?

A study of this question has been presented in a very elegant fashion by Feller in his book "Introduction to Probability Theory and its Applications." In the rest of this chapter we present some of the highpoints of this theory following Feller's treatment. The reader is warned that while the mathematics is not difficult it is tricky.

We begin by simulating a penny matching game between Mary and John. In Figure 11 we show the result of 100 matches.

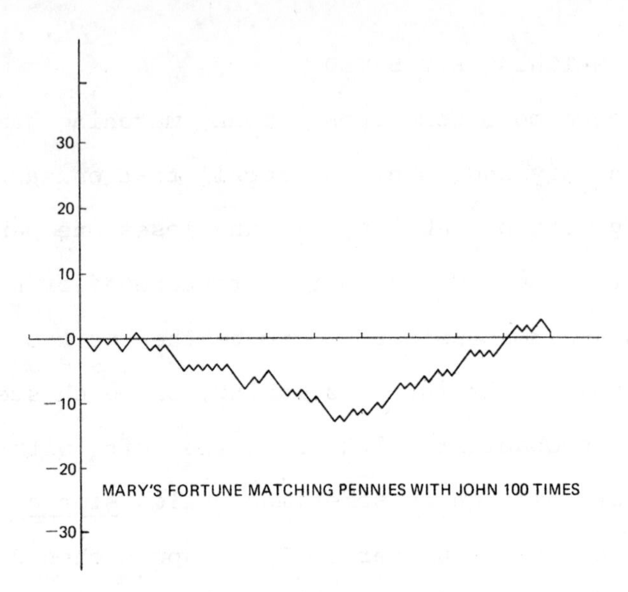

Figure 11.

We note that Mary had a lot of bad luck, but made a nice recovery. This raises the immediate question,"Can such a recovery be counted on?". That is, will Mary be sure of returning to 0 if she waits long enough no matter how badly she may start off. We can try a longer simulation to see if this suggests the answer. In Figure 12 we show the results of 10,000 matches.

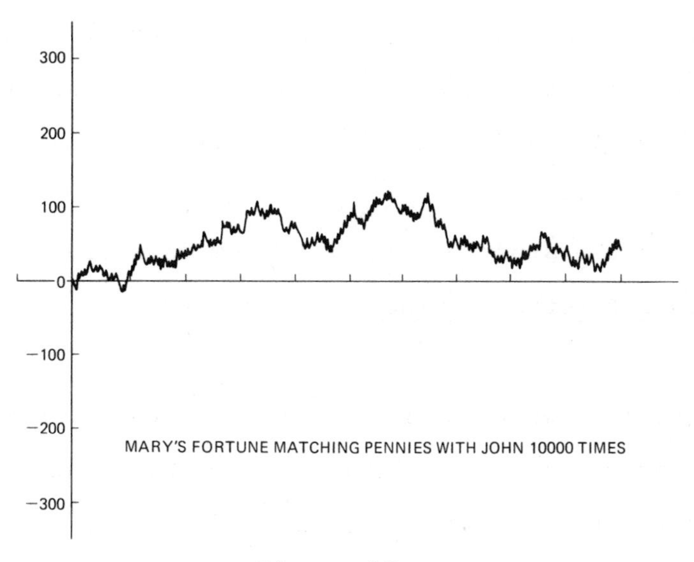

Figure 12.

We see that Mary's fortune did return to 0, but after that, she went in the lead and never returned to 0. It is intuitively clear that any time Mary's fortune is at 0 the behavior in the future should be the same as starting at 0. Thus we still are in the dark as to whether she can be confident of returning to 0 on any game. We discussed this question in Section 1 of Chapter 2, when we considered the question of the number of times a player would be in a lead. We did this only for a small number, 12, of plays by direct enumeration. We found some quite surprising results. Namely, that for 12 matches the most likely number of times in the lead is 0 or 12 and the least likely is 6. In our two simulations we note that in the first case Mary was behind most of the time and in the second she was ahead most of the time. This lends support to the theory that, in general, a large number of times in the lead, or a small number of times is more likely than one half the time. In the next section we shall study this problem and find an exact distribution for the number of times in the lead.

Basic to the study of this problem is a combinatorial trick called the reflection principle. We begin by explaining this principle.

In Figure 13 we have drawn a possible path for a random walk with 10 steps. The walk starts at a point A which is above the x-axis, hits the x-axis at B, and ends up at a point C again above the x-axis.

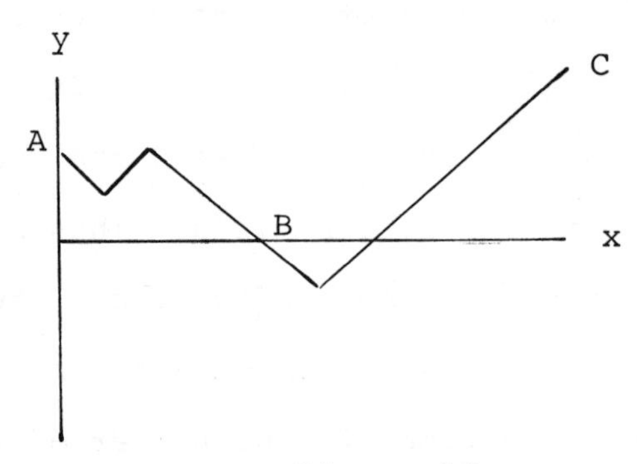

Figure 13.

Assume that we wish to count all possible paths from A to C which touch the x-axis at some point. Any path we wish to count has a segment (A,B) representing the outcomes until the first time the x-axis is reached. See Figure 13. We reflect this part of the path across the x-axis. We then obtain a path from $\bar{A}$ to C as indicated in Figure 14.

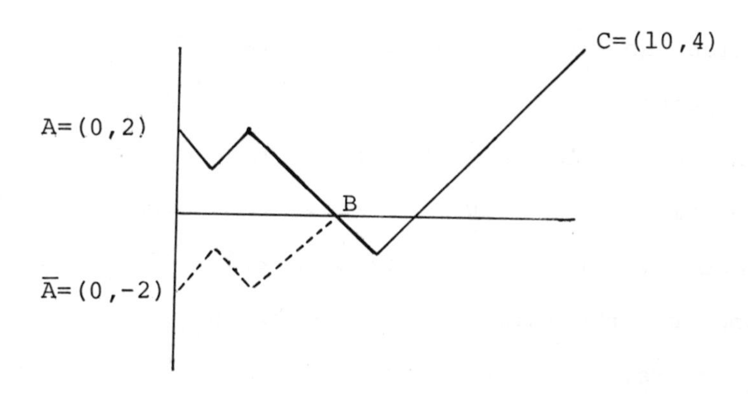

Figure 14.

Now every path from $\bar{A}$ to C crosses the x-axis. If we have any such path we can reflect the part $(\bar{A},B)$ corresponding to the first time the path touches the x-axis and obtain a path from A to B which touches the x-axis. Therefore there is a one-to-one correspondence between the number of paths from A to C which hit the x-axis and the total number of paths from $\bar{A}$ to C. This is the reflection principle.

Example. The coordinates of the points A and C in Figure 14 are (0,2) and (10,4). The coordinates of the point $\bar{A}$ are (0,-2). Consider a path from (0,-2) to (10,4). Let a be the number of times the path goes up and b the number of times it goes down. Then $a + b = 10$ and $a - b = 4 - (-2) = 6$. Thus $2a = 16$ or $a = 8$ and $b = 2$. The number of possible paths from $\bar{A}$ to C is then $C(10,8) = \frac{10!}{8!2!} = 45$. But then, by the reflection principle, this is also the number of paths from A to C which touch the x-axis. Assume now that we want to find the number of paths from

A to C which are always above the x-axis. We can find the total number of paths from A to C as before. Now a + b = 10 and a - b = 2. Thus a = 6, b = 4. Thus the number of paths from A to C with no restriction is C(10,6) = 210. The number of paths from A to C that do not intersect the axis is 210 - 45 = 165.

We now use the reflection principle to prove a fundamental theorem which relates a fact about the entire history of a random walk of length 2n to a fact about the position at time 2n. Recall that S_n can be 0 only at even number of plays. This enables us in many of our probabilities to consider only even n.

THEOREM. For a simple random walk

$$P(S_2 \neq 0,\ S_4 \neq 0, \ldots, S_{2n-2} \neq 0, S_{2n} \neq 0) = P(S_{2n} = 0) = \frac{1}{2^{2n}} \binom{2n}{n}.$$

Proof. Consider first, for the case r > 0,

$$(1) \qquad P(S_2 > 0,\ S_4 > 0,\ \ldots, S_{2n-2} > 0,\ S_{2n} = 2r).$$

For this event to occur we must (a), have $S_1(\omega) = 1$ and (b), go from (1,1) to (2n,2r) without hitting the x-axis. The probability of (a) is 1/2. Consider now (b). The probability that an unrestricted path from (1,1) ends at (2n,2r) is the same as the probability that a path from (0,0) ends at (2n-1,2r-1), i.e., $P(S_{2n-1} = 2r-1)$. To obtain the desired probability we subtract from this unrestricted probability, the probability that a path goes from (1,1) to (2n,2r) and hits the x-axis. By the reflection principle this probability is the same as the probability that a path from (1,-1) ends at (2n,2r). This in turn is the probability that an unrestricted path from (0,0) ends at (2n-1,2r+1), i.e., $P(S_n = 2r+1)$. Combining these results

$$P(S_2 > 0,\ S_4 > 0,\ldots,S_{2n-2} > 0, S_{2n} = 2r) \tag{2}$$

$$= \frac{1}{2}(P(S_{2n-1} = 2r-1) - P(S_{2n-1} = 2r+1)).$$

Consider now

$$\begin{aligned}
&P(S_2 > 0,\ S_4 > 0,\ldots,S_{2n-2} > 0,\ S_{2n} > 0)\\
&=P(S_2 > 0,\ S_4 > 0,\ldots,S_{2n-2} > 0,\ S_{2n} = 2)\\
&+P(S_2 > 0,\ S_4 > 0,\ldots,S_{2n-2} > 0,\ S_{2n} = 4)\\
&\quad\cdot\\
&\quad\cdot\\
&\quad\cdot\\
&+P(S_2 > 0,\ S_4 > 0,\ldots,S_{2n-2} > 0,\ S_{2n} = 2n-2)\\
&+P(S_2 > 0,\ S_4 > 0,\ldots,S_{2n-2} > 0,\ S_{2n} = 2n)
\end{aligned}$$

By (2) this is

$$\begin{aligned}
&\tfrac{1}{2}\,(P(S_{2n-1} = 1) - P(S_{2n-1} = 3)) \\
&+ \tfrac{1}{2}\,(P(S_{2n-1} = 3) - P(S_{2n-1} = 5)) \\
&\cdot \\
&\cdot \\
&\cdot \\
&+ \tfrac{1}{2}(P(S_{2n-1} - 2n-3) - P(S_{2n-1} = 2n-1)) \\
&+ \tfrac{1}{2}(P(S_{2n-1} = 2n-1) - P(S_{2n-1} = 2n+1))
\end{aligned}$$

All terms except the first and the last cancel. The last term is clearly 0 and hence

$$P(S_2 > 0,\ S_4 > 0,\ldots,S_{2n-2} > 0,\ S_{2n} > 0) = \tfrac{1}{2}\,P(S_{2n-1} = 1).$$

By symmetry

$$P(S_2 < 0,\ S_4 < 0,\ldots,S_{2n-2} < 0, S_{2n} < 0) = \tfrac{1}{2}\,P(S_{2n-1} = -1).$$

By adding these we obtain the desired result

$$P(S_2 \neq 0,\ S_4 \neq 0,\ldots,\ S_{2n-2} \neq 0, S_{2n} \neq 0) = P(S_{2n} = 0).$$

<u>Example</u>. In 12 matches the probability that one player is always in the lead is $P(S_{12} = 0) = B(12,6) \doteq .2256$. This agrees with the result obtained in Chapter 2, Section 1, by direct enumeration. Thus in 12 plays there is a quite reasonable chance that one player leads all the time. The probability that Mary leads all the time is, by symmetry, one half this value, or .1128.

We can use our theorem to answer the question: Will a random walk starting at 0 return to 0? Let T be the first time such a walk returns to 0. Then T is a random variable which depends upon the entire history and can take on any value 2,4,6,.... Fortunately, the probability that the first return to zero occurs at time 2k, $f_{2k} = P(T = 2k)$, depends only on the first 2k outcomes and so we can compute this.

We note next that

$$\begin{aligned}&P(S_2 \neq 0,\ S_4 \neq 0,\dots,S_{2k-2} \neq 0)\\&= P(S_2 \neq 0,\ S_4 \neq 0,\dots,\ S_{2k-2} \neq 0, S_{2k} \neq 0)\\&+ P(S_2 \neq 0,\ S_4 \neq 0,\dots,\ S_{2k-2} \neq 0,\ S_{2k} = 0).\end{aligned}$$

Thus

$$\begin{aligned}f_{2k} &= P(S_2 \neq 0,\ S_4 \neq 0,\dots,S_{2k-2} \neq 0)\\&- P(S_2 \neq 0,\ S_4 \neq 0,\dots,S_{2k-2} \neq 0,\ S_{2k} \neq 0).\end{aligned}$$

We now let $u_{2k} = P(S_{2k} = 0)$ with the convention that $u_0 = 1$. Then by our first theorem:

$$f_{2k} = u_{2k-2} - u_{2k},\ k = 1,2,\dots. \tag{3}$$

To find the probability that the walk returns to zero we look at the infinite series $f_2 + f_4 + f_6 + \dots$. We first note that

$$(4) \quad f_2 + f_4 + \ldots + f_{2n}$$
$$= u_0 - u_2 + u_2 - u_4 + \ldots . + u_{2n-2} - u_{2n}$$
$$= 1 - u_{2n}.$$

But we know that

$$u_{2n} \mp \binom{2n}{n} \frac{1}{2^{2n}} \mp \frac{1}{\sqrt{\pi n}}$$

and since this term tends to 0 as n tends to infinity, the infinite series (4) converges to 1. But this tells us that the probability that the walk returns to 0 for some 2k is equal to 1.

We can simplify our expression for f_{2k} as follows.

$$u_{2k-2} = \frac{1}{2^{2k-2}} \frac{(2k-2)!}{(k-1)!(k-1)!}$$
$$= \frac{1}{2^{2k-2}} \times \frac{2k!kk}{k!k!(2k)(2k-1)}$$
$$= u_{2k} \frac{4k^2}{2k(2k-1)} .$$

Thus

$$f_{2k} = u_{2k-2} - u_{2k} = u_{2k}\left(\frac{4k^2}{2k(2k-1)} - 1\right).$$
$$= u_{2k}\frac{1}{2k-1}$$

Consider now the expected value of T

$$E(T) = \sum_k 2kf_{2k}$$
$$= \sum_k \frac{2k}{2k-1} u_{2k}.$$

The kth term of this sum may be approximated by $\frac{1}{\sqrt{\pi k}}$. Thus the series diverges and the expected value of T is infinite.

Thus while return has probability one, the expected time to return is infinite. This means that we can expect long periods during which one player leads. This was borne out in our simulation. It also helps to explain the unusual nature of the distribution of the time one player is in the lead. We consider the problem in more detail in the next section.

EXERCISES

1. Mary and John match pennies 4 times. Find the probability that Mary is in the lead 0,2, and 4 times.
2. In 20 matches between Mary and John, what is the expected number of times Mary will be in the lead?
3. Find the probability that a random walk returns to the origin in less than or equal to 6 steps.
4. Let $A_{2k,2n}$ be the probability that in 2n steps of a random walk, the walk is at 0 for the last time at time 2k. Show that $A_{2k,2n} = u_{2k}u_{2n-2k}$.
5. Write a computer program to match pennies for 12 times and record the last time Z_{12} that the fortunes were equal in the 12 matches. Repeat this experiment 10,000 times to estimate the distribution of Z_{12}. Compare your estimates with (a) the exact values found in Exercise 4 and (b) the distribution for the number of times L_{12} in the lead found in Section 1 of Chapter 2.
6. On the basis of the results of Exercise 5, can you conjecture a relation between the distribution of L_{12} and Z_{12}?
7. In 2n steps of a random walk, which is more likely, that the last return to 0 will occur in the first n steps or in the last n steps?

6*. THE ARC SINE LAW

We return now to the study of the random variable $L_{2n}(\omega)$ representing the number of times Mary leads John in 2n matches or, equivalently, the number of times a random walk is positive in 2n steps. In this section we shall find the distribution of L_{2n}. Recall that this random variable can take on only even values. We define

$$v_{2n,2k} = P(L_{2n} = 2k).$$

We begin by proving the following theorem:

THEOREM. Let L_{2n} be the number of times in the first 2n steps a random walk is positive. Then

$$v_{2n,2k} = P(L_{2n} = 2k) = u_{2k}u_{2n-2k}$$

where $u_{2k} = P(S_{2k} = 0)$.

Proof. In the last section we proved, using the reflection principle, that

$$P(S_2 > 0,\ S_4 > 0, \ldots,\ S_{2n} > 0) = \frac{1}{2}\, u_{2n}.$$

This event may be regarded as (a) the walk going up on the first step (probability 1/2), and (b) starting from 1 remaining ≥ 1 on the next 2n-1 steps. But (b) is equivalent to the process started at 0 remaining ≥ 0 for 2n-1 steps. We can even require that it remain ≥ 0 for 2n steps since S_{2n-1} cannot be 0. Thus

$$P(S_2 > 0,\ S_4 > 0,\ldots,S_{2n-2} > 0,\ S_{2n} > 0)$$
$$= \frac{1}{2}\,P(S_2 \geq 0, S_4 \geq 0,\ldots,S_{2n-2} \geq 0,\ S_{2n} \geq 0) = \frac{1}{2}\,u_{2n}$$

or

$$v_{2n,2n} = P(S_2 \geq 0,\ S_4 \geq 0,\ldots,\ S_{2n} \geq 0) = u_{2n}\ .$$

By symmetry $v_{2n,0} = u_{2n}$. Thus for any n the theorem is true for $k = 0$ and n. For $n = 1$, $v_{2,0} = v_{2,2} = \frac{1}{2}$. On the other hand, $u_0 = 1$ and $u_2 = \frac{1}{2}$. Thus the theorem is true for $n = 1$. We shall now prove the theorem by induction. Assume that the theorem is true for $m \leq n-1$. We shall prove it is true for n. We need only consider $1 \leq k \leq n-1$. For such a k, for the paths which contribute to the probability $v_{2n,2k}$ there must be a first time that 0 is reached after time 0. Let this time be 2r. Before this time there must have been either a negative or a positive period. Taking these two alternatives separately we have

$$v_{2n,2k} = \frac{1}{2}\sum_{r=1}^{k} f_{2r}v_{2n-2r,2k-2r}$$
$$+ \frac{1}{2}\sum_{r=1}^{n-k} f_{2r}v_{2n-2r,2k}.$$

By the induction hypothesis this is

$$= \frac{1}{2} \sum_{r=1}^{k} f_{2r}u_{2k-2r}u_{2n-2k} + \frac{1}{2} \sum_{r=1}^{n-k} f_{2r}u_{2k}u_{2n-2k-2r}$$

$$= \frac{1}{2} u_{2n-2k} \sum_{r=1}^{k} f_{2r}u_{2k-2r} + \frac{1}{2} u_{2k} \sum_{r=1}^{n-k} f_{2r}u_{2n-2r-2k}$$

$$= \frac{1}{2} u_{2n-2k}u_{2k} + \frac{1}{2} u_{2n-2k}u_{2k}$$

$$= u_{2k}u_{2n-2k}$$

as was to be proven.

We now have a simple formula for the distribution of L_{2n}, namely,

$$P(L_{2n} = 2k) = \binom{2k}{k}\binom{2n-2k}{n-k} \frac{1}{2^{2n}} .$$

Example. We computed in Chapter 2 by direct enumeration the distribution of L_{12}. We can easily modify the program BINOM to compute the distribution of L_{12} using the formula just derived. The Program LEAD1 is such a modification.

```
LEAD1

100 DEF FNF(X) = INT(1000*X+.5)/1000
110 DIM B(50,50)
120 FOR N = 0 TO 50
130    LET B(N,0) = B(N,N) = 1/(2↑N)
140    FOR J = 1 TO N-1
150       LET B(N,J) = .5*B(N-1,J-1) + .5*B(N-1,J)
160    NEXT J
170 NEXT N
180 FOR I = 0 TO 6
190    LET X = B(2*I,I)*B(12-2*I,6-I)
200    PRINT 2*I,FNF(X)
210 NEXT I
220 END
```

```
LEAD1

0              0.226
2              0.123
4              0.103
6              0.098
8              0.103
10             0.123
12             0.226
```

The reader may check that the values computed by our formula agree with those obtained in Chapter 2 by direct enumeration.

We are now in a position to look at the distribution of L_{2n} for larger values of n and to see if, as in the case of sums of independent random variables, we have a limiting behaviour.

In Figure 15 we have graphed the distribution for the number of times in the lead for n = 12,20, and 50.

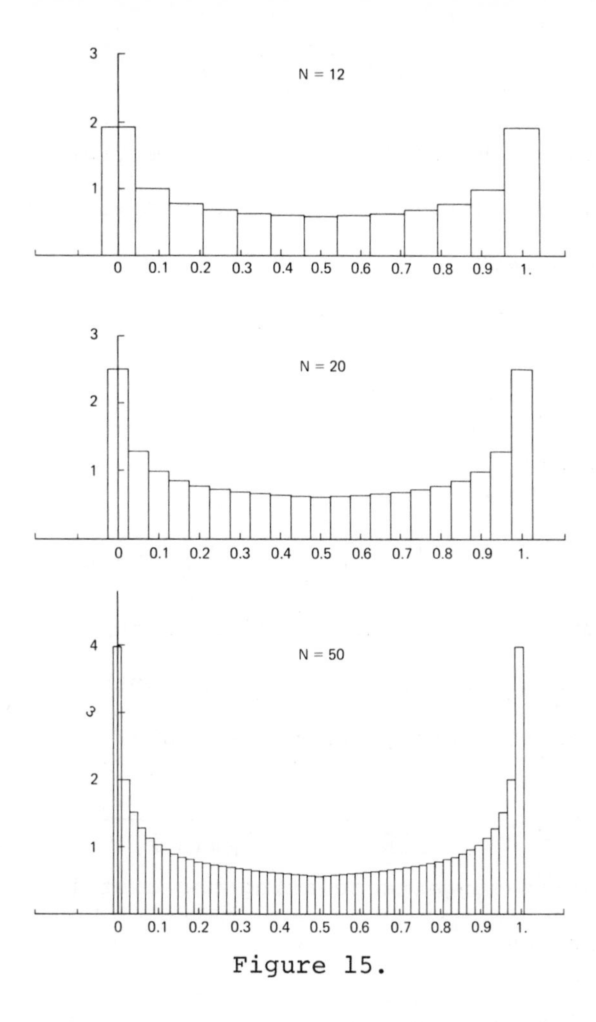

Figure 15.

As usual we have graphed the distribution as a histogram. That is, the probability of a specific outcome is represented by the area of a rectangle. More specifically, the area of the kth rectangle represents the $P(L_{2n} = 2k)$.

We note that the shape of the distributions are quite similar. We continue to have the extreme values most likely and the central values the least likely. As is suggested by the graphs there is a limiting graph and we now give an indication of the proof of this. We shall see that the limiting distribution in this case is very different from the normal distribution found for independent trials.

We have used many times the estimate

$$u_{2n} \doteqdot \frac{1}{\sqrt{\pi n}} .$$

Consider then

$$P(L_{2n} = 2k) = v_{2n,2k} = u_{2k}u_{2n-2k}.$$

We shall examine these probabilities when n and k tend to infinity in such a way that n and n-k both tend to infinity. Then using our approximation for u_{2n} we have

$$v_{2n,2k} \doteqdot \frac{1}{\sqrt{\pi k}} \cdot \frac{1}{\sqrt{\pi(n-k)}} = \frac{1}{\pi\sqrt{k(n-k)}} .$$

This approximation suggests relating the distribution of L_{2n} to the curve defined by

$$f(x) = \frac{1}{\pi\sqrt{x(1-x)}}$$

defined for $0 < x < 1$. A graph of this function is given in Figure 16.

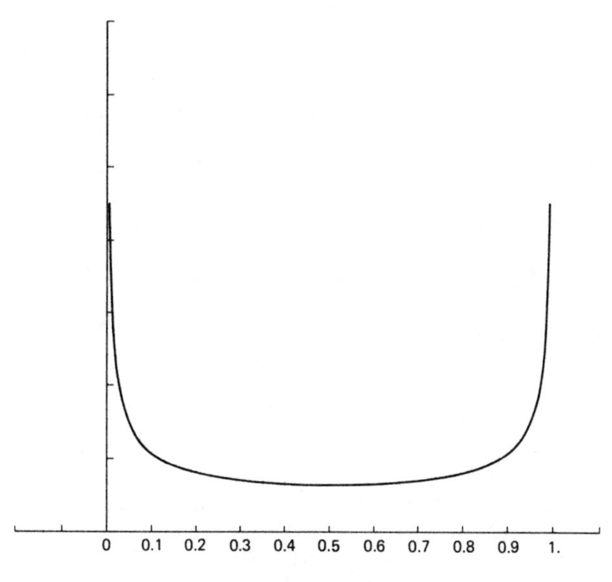

Figure 16.

As in the case of the normal approximation, we shall consider the area under this curve. Let A(a,b) be the area between a and b. If this area is to be well defined we would obtain it from a limiting process of find the areas of approximating rectangles.

We proceed by dividing the unit interval into subintervals of length 1/n. We construct n approximating rectangles as follows. For the kth rectangle we use the kth interval as base and the value of the function f(x) at the midpoint of the interval (k/n,(k+1)/n) as height. The approximation $A_n(a,b)$ for the area between a and b then, is the sum of the areas of the rectangles for k such that

$$a < k/n < b$$

In Figure 17 we show the result of approximating the area between .1 and .4 by taking subintervals of length .01.

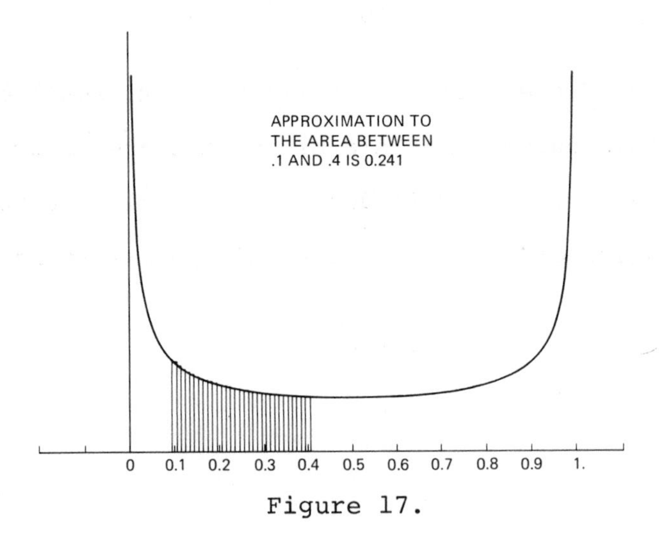

Figure 17.

We next consider the connection between this area and the distribution of L_{2n}. Let us rewrite our approximation (1) in the form

$$v_{2n,2k} \doteq \frac{1}{n} \cdot \frac{1}{\pi\sqrt{\frac{k}{n} \cdot (1 - \frac{k}{n})}} .$$

We note that this approximation for $v_{2n,2k}$ is precisely the area of the kth rectangle used in finding the area under part of the curve defined by f(x). Now

$$P(a < \frac{L_{2n}}{2n} < b) = \sum_{a < k/n < b} v_{2n,2k} .$$

This suggests the approximation

$$P(a < \frac{L_{2n}}{2n} < b) \doteq A_n(a,b)$$

These approximations can be shown to hold in the limit and the following limit theorem is obtained.

THEOREM. (Arc sine law). Let L_{2n} be the number of times a player is in the lead in 2n penny matches. Then as n tends to infinity

$$P(a < \frac{L_{2n}}{2n} < b) \dashrightarrow A(a,b)$$

where A(a,b) is the area under the curve defined by

$$f(x) = \frac{1}{\pi\sqrt{x(1-x)}}$$

between a and b. Figure 18 shows the comparison of the distribution of L_{50} and the curve determined by f(x).

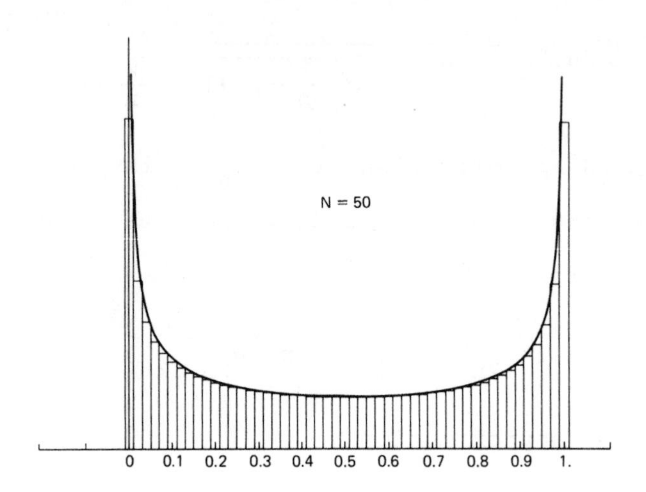

Figure 18.

Unlike the normal function, we can find, by the methods of calculus, a formula for the area A(a,b). This is because the function

$$h(x) = \frac{2}{\pi} \text{ arc sin } \sqrt{x}$$

has as its derivative f(x). This also shows that the area between 0 and x is well defined and enables us to conclude that

$$P\left(\frac{L_{2n}}{2n} < x\right) \rightarrow \frac{2}{\pi} \text{ arc sin } \sqrt{x}.$$

The following table gives the exact values for

$$P\left(\frac{L_{1000}}{1000} < x\right).$$

and the approximation given by $\frac{2}{\pi}$ arc sin $\sqrt{x}$.

x	$P(\frac{L_{1000}}{1000} < x)$	$\frac{2}{\pi}$ arc sin $\sqrt{x}$
0	.025	0
.02	.094	.090
.04	.131	.128
.06	.159	.158
.08	.184	.183
.10	.206	.205
.12	.227	.225
.14	.245	.244
.16	.263	.262
.18	.280	.279
.20	.296	.295
.22	.312	.311
.24	.327	.326
.26	.342	.348
.28	.356	.355
.30	.370	.369
.32	.384	.383
.34	.397	.396
.36	.410	.410
.38	.424	.423
.40	.430	.429
.42	.450	.449
.44	.462	.462
.46	.475	.474
.48	.488	.481
.50	.500	.500

We see that the approximations are quite good.

<u>Example</u>. We note that in 1000 matches Mary leads $\leq$ 20 times with probability .094. By symmetry she leads 980 or more times with the same probability. Thus, in about 20 percent of our penny matching games we would expect one player to be in the lead 98 percent of the time. This reflects the fact that very long

times in the lead for one player are to be expected as indicated in our simulations.

EXERCISES

1. In 1000 penny matches, what is the probability that Peter will be in the lead less than 5 percent of the time?
2. In 1000 penny matches, what is the probability that the fraction of time Peter will be in the lead will be between .45 and .5?
3. Simulate 20 coin matches by tossing a coin 20 times and graph the outcomes.
4. Assume the model of a simple random walk for the price of a stock. That is, each day it goes up \$1 with probability $\frac{1}{2}$ and down \$1 with probability $\frac{1}{2}$. Assume that there are 196 trading days in a year.

   (a) Estimate the probability that the stock has increased in value by at least \$14 at the end of the year.

   (b) On a day that the price is above the price he paid for it, Jones is in a good mood. Estimate the probability that he is in a good mood at least 3/4 of the time.
5. (a) For $r \leq x$, $x = 0, 1, \ldots$, determine $P(S_n = r \text{ and } \max_{j \leq n} S_j = x)$ (Hint: Use the reflection principle about $y = x$.)

   (b) Use part (a) to show that

$$P(\max_{j \leq n} S_j = x) = \begin{cases} P(S_n = x) & \text{for } n-x \text{ even} \\ P(S_n = x+1) & \text{for } n-x \text{ odd.} \end{cases}$$

6. From Exercise 5, obtain for $r > 0$ an explicit evaluation of

$$P(S_n = r;\ S_j < r \text{ for } j < n).$$

   This is known as the _first passage_ probability through r at time n.

7. Use computer simulations to estimate the distribution of

$$P(L_{2n} = 2k \text{ and } S_{2n} = 0) \text{ for } n = 10 \ (k = 0,1,\ldots,n).$$

   (As less than 1 in 5 trials will have $S_{20} = 0$, a large number of experiments will be required.)

CHAPTER 4

MARKOV CHAINS

1. INTRODUCTION

We have so far studied only one kind of stochastic process, namely independent trials processes. That is, a process such that the outcomes $X_1(\omega)$, $X_2(\omega),\ldots,X_n(\omega)$ are independent and have the same distribution. For such a process the conditional probability

$$P(X_n = t|X_{n-1} = s, X_{n-2} = r,\ldots,X_1 = a) = p_t$$

does not depend on the outcomes $a,\ldots,r,s$ of the first $n-1$ experiments or on the number n of the experiment being considered. The probabilities $P(X_k = a) = p_a$ are the same for all $k = 1,2,\ldots,n$ and these basic probabilities determine all probabilities relating to the sequence of experiments. We consider now the next more complicated situation when we allow the probabilities of the outcome of the nth experiment to depend upon the outcome of the previous experiment. That is,

$$P(X_n(\omega) = t|X_{n-1}(\omega) = s,\ X_{n-2}(\omega) = r,\ldots,X_1(\omega) = a) = p_{st}$$

does not depend upon the values $a,b,\ldots,r$ or on n, only on s and t. Such a process is called a <u>Markov chain</u> and may be described more formally as follows.

DEFINITION. A <u>Markov chain</u> is determined by specifying the following information: There is given a set of <u>states</u> $S = \{s_1, s_2, \ldots, s_r\}$. The process can be in one, and only one, of these states at a given time, and it moves successively from one state to another. Each move is called a <u>step</u>. The probability that the process moves from s_i to s_j depends only on the state s_i that it occupied before the step. A <u>transition probability</u> p_{ij}, which gives the probability that the process will move from s_i to s_j, is given for every pair of states. An <u>initial distribution</u> defined on S specifies how the process is started. This will usually specify a particular state as the starting state.

With the information given above it is now possible to construct a tree and assign path weights to describe the Markov chain process when it moves through any (finite) number of steps.

<u>Example 1</u>. According to Kemeny, Snell, and Thompson, "Finite Mathematics," the Land of Oz is blessed by many things, but not by good weather. They never have two nice days in a row. If they have a nice day they are just as likely to have snow as rain the next day. If they have snow or rain, they have an even chance of having the same the next day. If there is change from snow or rain, only half of the time is this a change to a nice day. It is a nice day today in the Land of Oz. With this information we form a Markov chain as follows. We take as states the kinds of weather R, N, S. From the above information we

determine the transition probabilities. These are most conveniently represented in a square array as

$$\begin{array}{c} \\ R \\ N \\ S \end{array} \begin{array}{c} \begin{array}{ccc} R & N & S \end{array} \\ \begin{pmatrix} 1/2 & 1/4 & 1/4 \\ 1/2 & 0 & 1/2 \\ 1/4 & 1/4 & 1/2 \end{pmatrix} \end{array} .$$

The entries in the first row represent the probabilities for the various kinds of weather following a rainy day, those in the second row represent these probabilities following a nice day, and the same for a snowy day in the third row. Such a square array is called the <u>matrix of transition probabilities</u>. From this we determine a tree and a tree measure for the next three days weather as indicated in Figure 1.

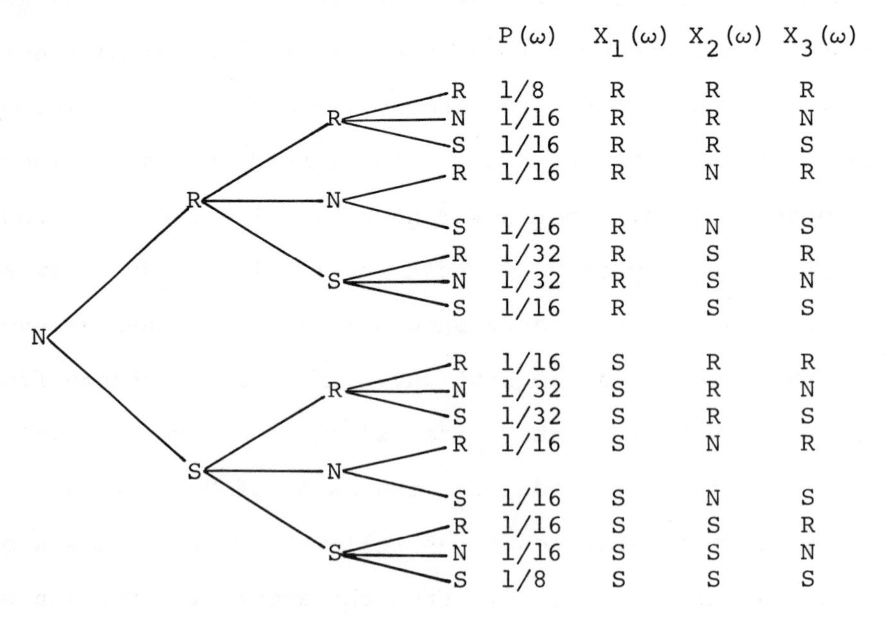

$P(\omega)$	$X_1(\omega)$	$X_2(\omega)$	$X_3(\omega)$
1/8	R	R	R
1/16	R	R	N
1/16	R	R	S
1/16	R	N	R
1/16	R	N	S
1/32	R	S	R
1/32	R	S	N
1/16	R	S	S
1/16	S	R	R
1/32	S	R	N
1/32	S	R	S
1/16	S	N	R
1/16	S	N	S
1/16	S	S	R
1/16	S	S	N
1/8	S	S	S

Figure 1.

The sample space Ω consists of the paths of the tree. The tree measure $P(\omega)$ is assigned in the usual manner. The random variables X_1, X_2, X_3 then give, respectively, the outcome for the first, second, and third days' weather. From the tree measure we can find the distribution of these three random variables. For Example, X_1 has the distribution,

$$P(X_1 = R) = 1/2$$
$$P(X_1 = N) = 0$$
$$P(X_1 = S) = 1/2$$

We shall denote by $w^{(n)}$ the distribution of X_n and exhibit these distributions by row vectors,

$$w^{(n)} = (w_1^{(n)}, w_2^{(n)}, \ldots, w_r^{(n)})$$

where the jth component refers to state s_j. From the tree measure we obtain,

$$w^{(1)} = (1/2 \quad 0 \quad 1/2)$$
$$w^{(2)} = (3/8 \quad 2/8 \quad 3/8)$$
$$w^{(3)} = (13/32 \quad 6/32 \quad 13/32).$$

In studying independent processes we found it necessary to develop methods for computing properties of the process without resorting to the tree and the tree measures. The same is true for the case of Markov chains. In particular, we shall now show how to obtain the vectors $w^{(n)}$ directly from the transition matrix. The key to our method is the following simple observation.

$$w_j^{(n+1)} = P(X_{n+1} = s_j)$$

$$= \sum_k P(X_n = s_k, X_{n+1} = s_j)$$

$$= \sum_k P(X_n = s_k) P(X_{n+1} = s_j | X_n = s_k).$$

For a Markov chain process, the second term in each product does not depend on n and is, in fact, p_{kj}. Therefore

$$w_j^{(n+1)} = \sum_k w_k^{(n)} p_{kj}.$$

In matrix language this is

$$W^{(n+1)} = W^{(n)} P.$$

Thus we see that the vector giving the probabilities of being in each of the states after $n+1$ steps is obtained by multiplying the transition matrix on the left by the probability vector corresponding to n steps. Thus, if we start in s_i, $W^{(1)}$ is the ith row of P, $W^{(2)} = W^{(1)}P$, $W^{(3)} = W^{(2)}P$, etc.

Example 1. (Continued). In the Land of Oz example, let us start in state N. Then $W^{(1)} = (1/2 \ 0 \ 1/2)$,

$$W^{(2)} = (1/2 \ \ 0 \ \ 1/2)\begin{pmatrix} 1/2 & 1/4 & 1/4 \\ 1/2 & 0 & 1/2 \\ 1/4 & 1/4 & 1/2 \end{pmatrix}$$

$$= (3/8 \ \ 2/8 \ \ 3/8)$$

$$W^{(3)} = (3/8 \ \ 2/8 \ \ 3/8)\begin{pmatrix} 1/2 & 1/4 & 1/4 \\ 1/2 & 0 & 1/4 \\ 1/4 & 1/4 & 1/2 \end{pmatrix}$$

$$= (13/32 \ \ 6/32 \ \ 13/32).$$

Note that these vectors agree with those calculated earlier from the tree and its tree measure.

Observe that if we start in s_i, $W^{(1)}$ is the ith row of P, $W^{(2)} = W^{(1)}P$ is the ith row of P^2, etc. Hence $W^{(n)}$ is the ith row of P^n. Thus the various rows of P^n give us $W^{(n)}$ for various starting states. Let $p_{ij}^{(n)}$ be the probability that the process will be in state s_j after n steps if it started in state s_i. It is the ijth entry of the matrix P^n.

<u>Example 1</u>. (Continued). Let us consider again the Land of Oz example. We have just seen that the powers of the transition matrix give us interesting information about the process as it evolves. We shall be particularly interested in the position of the chain after a large number of steps. The program POWER computes the powers of P for powers of the form 2^m, m = 1,2,..,N. We show a run of this program in N = 3 in our example.

```
POWER

100 READ N
110 MAT READ P(N,N)
120 PRINT "TRANSITION MATRIX P"
130 MAT PRINT P
140 PRINT
150 FOR T = 1 TO 3
160    PRINT "MATRIX P TO POWER";2↑T
170    MAT P = P*P
180    MAT PRINT P
190    PRINT
200 NEXT T
210 DATA 3
220 DATA .5,.25,.25
230 DATA .5,0,.5
240 DATA .25,.25,.5
250 END
```

```
POWER

TRANSITION MATRIX P

.5             .25            .25
.5             0               .5
.25            .25            .5

MATRIX P TO POWER 2

.4375          .1875          .375
.375           .25            .375
.375           .1875          .4375

MATRIX P TO POWER 4

.402344        .199219        .398437
.398437        .203125        .398437
.398437        .199219        .402344

MATRIX P TO POWER 8

.400009        .199997        .399994
.399994        .200012        .399994
.399994        .199997        .400009
```

We note that after 8 days our weather predictions would be essentially independent of today's weather. The probabilities for the three types of weather, R, N, and S, are very near to .4,.2,.4 for each of the possible starting states. This example is an example of a type of Markov chain called a <u>regular Markov chain</u>. For this type of chain it is true that long range predictions are independent of the starting state. Not all chains are regular, but this is an important class that we shall study in detail later.

The following examples of Markov chains will be used throughout the chapter for Exercises.

<u>Example 2</u>. The President of the United States tells person A his intention either to run or not to run in the next election. Then A relays the news to B, who in turn relays the message to C, etc., always to some new person. We assume that there is a probability a that a person will change the answer from "yes" to "no" when transmitting it to the next person and a probability b that he will change it from "no" to "yes." We choose as states the message, either "yes" or "no." The transition matrix is then

$$\begin{array}{c} \\ \text{yes} \\ \text{no} \end{array} \begin{array}{c} \begin{array}{cc} \text{yes} & \text{no} \end{array} \\ \begin{pmatrix} 1-a & a \\ b & 1-b \end{pmatrix} \end{array}.$$

The initial state represents the President's choice.

<u>Example 3</u>. Each time a certain horse runs a race she has probability 2/5 of winning, 1/5 of tying, and 2/5 of losing,

independent of the outcome of any previous race. We have here an independent trials process, but it may also be considered from the point of view of Markov chain theory. The transition matrix is

$$\begin{array}{c} \\ W \\ T \\ L \end{array} \begin{array}{c} \begin{array}{ccc} W & T & L \end{array} \\ \begin{pmatrix} .4 & .2 & .4 \\ .4 & .2 & .4 \\ .4 & .2 & .4 \end{pmatrix} \end{array} .$$

<u>Example 4</u>. Assume that of the sons of Harvard men, 80 percent go to Harvard and the rest go to Yale, 40 percent of the sons of Yale men go to Yale, and the remaining split evenly between Harvard and Dartmouth; and of the sons of Dartmouth men, 70 percent go to Dartmouth, 20 percent go to Harvard, and 10 percent go to Yale. We form a Markov chain with transition matrix,

$$\begin{array}{c} \\ H \\ Y \\ D \end{array} \begin{array}{c} \begin{array}{ccc} H & Y & D \end{array} \\ \begin{pmatrix} .8 & .2 & 0 \\ .3 & .4 & .3 \\ .2 & .1 & .7 \end{pmatrix} \end{array} .$$

<u>Example 5</u>. Modify Example 4 by assuming that the son of a Harvard man always goes to Harvard. The transition matrix now is:

$$
\begin{array}{c c}
 & \begin{array}{ccc} H & Y & D \end{array} \\
\begin{array}{c} H \\ Y \\ D \end{array} &
\begin{pmatrix} 1 & 0 & 0 \\ .3 & .4 & .3 \\ .2 & .1 & .7 \end{pmatrix}
\end{array} .
$$

<u>Example 6</u>. The following is a special case of a model which has been used to explain diffusion of gases. The general model will be discussed in detail in a later section. A box contains three balls. It has two compartments. Each second, one of the three balls is chosen at random and moved from its compartment into the other. We choose as state the number of balls in the first compartment. The transition matrix is then

$$
\begin{array}{c c}
 & \begin{array}{cccc} 0 & 1 & 2 & 3 \end{array} \\
\begin{array}{c} 0 \\ 1 \\ 2 \\ 3 \end{array} &
\begin{pmatrix} 0 & 1 & 0 & 0 \\ 1/3 & 0 & 2/3 & 0 \\ 0 & 2/3 & 0 & 1/3 \\ 0 & 0 & 1 & 0 \end{pmatrix}
\end{array} .
$$

<u>Example 7</u>. The simplest type of inheritance of traits in animals occurs when a trait is governed by a pair of genes, each of which may be of two types, say G and g. An individual may have a GG combination or Gg (which is genetically the same as gG) or gg. Very often the GG and Gg types are indistinguishable in appearance, and then we say that the G gene <u>dominates</u> the g gene. An individual is called <u>dominant</u> if he has GG genes, <u>recessive</u> if he has gg, and <u>hybrid</u> with a Gg mixture.

In the mating of two animals, the offspring inherits one gene of the pair from each parent, and the basic assumption of genetics is that these genes are selected at random, independently of each other. This assumption determines the probability of occurrence of each type of offspring. The offspring of two purely dominant parents must be dominant, of two recessive parents must be recessive, and of one dominant and one recessive parent must be hybrid. In the mating of a dominant and a hybrid animal, the offspring must get a G gene from the former and has probability $\frac{1}{2}$ for getting G or g from the latter, hence the probabilities are even of getting a dominant or a hybrid offspring. Again, in the mating of a recessive and a hybrid, there is an even chance of getting either a recessive or a hybrid. In the mating of two hybrids, the offspring has probability $\frac{1}{2}$ of getting G or a g from each parent. Hence the probabilities are 1/4 for GG, 1/2 for Gg, and 1/4 for gg.

Let us consider a process of continued crossings. We start with an individual of unknown genetic character, and cross it with a hybrid. The offspring is again crossed with a hybrid, etc. The resulting process is a Markov chain. The states are dominant, hybrid, and recessive, and indicated by D, H, and R, respectively. The transition probabilities are

$$\begin{array}{c} \\ D \\ H \\ R \end{array}\begin{array}{c} \begin{array}{ccc} D & \quad H & \quad R \end{array} \\ \begin{pmatrix} .5 & .5 & 0 \\ .25 & .5 & .25 \\ 0 & .5 & .5 \end{pmatrix} \end{array} .$$

Example 8. Now modify Example 7 so that we keep crossing the offspring with a dominant animal. The transition matrix is

$$\begin{array}{c c} & \begin{array}{ccc} D & H & R \end{array} \\ \begin{array}{c} D \\ H \\ R \end{array} & \begin{pmatrix} 1 & 0 & 0 \\ .5 & .5 & 0 \\ 0 & 1 & 0 \end{pmatrix} \end{array}.$$

Example 9. We start with two animals of opposite sex, cross them, select two of their offspring of opposite sex, and cross those, etc. To simplify the example, we will assume that the trait under consideration is independent of sex.

Here a state is determined by a pair of animals. Hence the states of our process will be: s_1 = (D,D), s_2 = (D,H), s_3 = (D,R), s_4 = (H,H), s_5 = (H,R), and s_6 = (R,R). Let us illustrate the calculation of transition probabilities in terms of s_2. When the process is in this state, one parent has GG genes, the other Gg. Hence the probability of a dominant offspring is $\frac{1}{2}$. Then the probability of transition to s_1 (selection of two dominants) is $\frac{1}{4}$, transition to s_2 is $\frac{1}{2}$, and to s_4 is $\frac{1}{4}$. The transition matrix of this chain and the powers P^2, P^4, P^8 of P are given by the following run of POWER.

```
POWER

TRANSITION MATRIX P

1.0000    .0000    .0000    .0000    .0000    .0000
 .2500    .5000    .0000    .2500    .0000    .0000
 .0000    .0000    .0000   1.0000    .0000    .0000
 .0625    .2500    .1250    .2500    .2500    .0625
 .0000    .0000    .0000    .2500    .5000    .2500
 .0000    .0000    .0000    .0000    .0000   1.0000

MATRIX P TO POWER 2

1.0000    .0000    .0000    .0000    .0000    .0000
 .3906    .3125    .0312    .1875    .0625    .0156
 .0625    .2500    .1250    .2500    .2500    .0625
 .1406    .1875    .0312    .3125    .1875    .1406
 .0156    .0625    .0312    .1875    .3125    .3906
 .0000    .0000    .0000    .0000    .0000   1.0000

MATRIX P TO POWER 4

1.0000    .0000    .0000    .0000    .0000    .0000
 .5420    .1445    .0215    .1367    .0820    .0732
 .2070    .1719    .0391    .2031    .1719    .2070
 .2627    .1367    .0254    .1758    .1367    .2627
 .0732    .0820    .0215    .1367    .1445    .5420
 .0000    .0000    .0000    .0000    .0000   1.0000

MATRIX P TO POWER 8

1.0000    .0000    .0000    .0000    .0000    .0000
 .6667    .0500    .0092    .0594    .0461    .1687
 .3742    .0734    .0141    .0906    .0734    .3742
 .3982    .0594    .0113    .0734    .0594    .3982
 .1687    .0461    .0092    .0594    .0500    .6667
 .0000    .0000    .0000    .0000    .0000   1.0000
```

We note that in this example, there is a high probability after eight steps to be in either the first state (D,D) or the last state (R,R). These states are traps which have the property that if we ever enter one of these states we cannot leave the state. Further, from any one state we can reach one of these traps. This is a special case of a type of chain that we shall call <u>absorbing</u>. We shall study such chains in detail later.

EXERCISES

1. It is raining in the Land of Oz. Determine a tree and tree measure for the next three days' weather. Find $W^{(1)}$, $W^{(2)}$, and $W^{(3)}$ and compare with the results obtained from P, P^2, and P^3.

2. In Example 2, let $a = 0$ and $b = 1/2$. Find P, P^2, and P^3. What would P^n be? What happens to P^n as n tends to infinity? Interpret this result.

   Ans. After a long time the answer will be "yes."

3. In Example 3, find P^2, and P^3. What is P^n?

   Ans. $P^n = P$.

4. For Example 4, find the probability that the grandson of a man from Harvard goes to Harvard.

5. In Example 5, find the probability that the grandson of a man from Harvard goes to Harvard.

   Ans. The probability is 1.

6. In Example 7, assume that we start with a hybrid. Find $W^{(1)}$, $W^{(2)}$, $W^{(3)}$. What would $W^{(n)}$ be?

   Ans. $W^{(n)} = W^{(1)}$.

7. Find the matrices P^2, P^3, P^4, and P^n for the Markov chain determined by the transition matrix $P = \begin{pmatrix} 1 & 0 \\ 0 & 1 \end{pmatrix}$. Do the same for the transition matrix $P = \begin{pmatrix} 0 & 1 \\ 1 & 0 \end{pmatrix}$. Interpret what happens in each of these processes.

8. A certain calculating machine uses only the digits 0 and 1. It is supposed to transmit one of these digits through several stages. However, at every stage there is a probability p that the digit that enters this stage will be changed when it leaves and hence probability $q = 1 - p$ of being transmitted unchanged. We form a Markov chain to represent the process of transmission by taking as states the digits 0 and 1. What is the matrix of transition probabilities?

9. For the Markov chain in Exercise 8, draw a tree and assign a tree measure, assuming that the process begins in state 0 and moves through three stages of transmission. What is the probability that the machine after three stages produces the digit 0, i.e., the correct digit? What is the probability that the machine never changed the digit from 0?

Ans. $3p^2q + q^3$; q^3.

10. Assume that a man's profession can be classified as professional, skilled laborer, or unskilled laborer. Assume that of the sons of professional men, 80 percent are professional, 10 percent are skilled laborers, and 10 percent are unskilled laborers. In the case of sons of skilled laborers, 60 percent are skilled laborers, 20 percent are professional, and 20 percent are unskilled. Finally, in the case of unskilled laborers, 50 percent of the sons are unskilled laborers, and 25 percent each are in

the other two categories. Assume that every man has a son, and form a Markov chain by following a given family through several generations. Set up the matrix of transition probabilities. Find the probability that the grandson of an unskilled laborer is a professional man.

Ans. .375

11. In Exercise 10 we assumed that every man has a son. Assume instead that the probability a man has a son is .8. Form a Markov chain with four states. The first three states are as in Exercise 10, and the fourth state is such that the process enters it if a man has no son, and that the state cannot be left. This state represents families whose male line has died out. Find the matrix of transition probabilities and find the probability that an unskilled laborer has a grandson who is a professional man.

Ans. .24.

12. Explain why the transition probabilities given in Example 6 are correct.

13. Write a program to compute $W^{(n)}$ given $W^{(1)}$ and P. Use this program to compute $W^{(10)}$ for the Land of Oz Example with $W^{(1)} = (1/3\ 1/3\ 1/3)$.

14. Using the program POWER find P^2, P^4, P^8 for Examples 7 and 8. See if you can predict the long range probability for finding the process in each of the states for these examples.

15. Modify the program POWER so that it prints out every power of P from 1 to 8. Use this to compute the first 8 powers of P for Example 6. Can you predict the nature of the powers of P for large N?

16. Modify the program POWER so that it prints out the average of the powers of the transition matrix. That is the average

$$A_n = (I + P + P^2 + \dots + P^{n-1})/n.$$

Compute these averages for n = 1 to 8 for Example 6. Can you predict the nature of these averages for large N?

17. Write a program to simulate the outcomes of a Markov chain given the intial starting state and the transition matrix P as data. Keep this program for use in later problems.

18. Modify the program of Exercise 17 so that it keeps track of the proportion of times in each state in n steps. Run the modified program for different starting states for the examples of the weather in the Land of Oz and also Example 6. Does the initial state effect the proportion of time spent in each of the states?

2. ABSORBING MARKOV CHAINS

The subject of Markov chains is best studied by considering special types of Markov chains. The first type that we shall study is called an <u>absorbing Markov chain</u>.

DEFINITION. A state of a Markov chain is called <u>absorbing</u> if it is impossible to leave it. A Markov chain is <u>absorbing</u> if (1) it has at least one absorbing state, and (2) from every state it is possible to go to an absorbing state (not necessarily in one step).

<u>Example</u>. A particle moves on a line; each time it moves one unit to the right or to the left with an equal probability of 1/2 in each case. We introduce barriers so that if it ever reaches one of these barriers it stays there, i.e., is absorbed. As a simple example, let the states be 0, 1, 2, 3, 4. States 0 and 4 are absorbing states. The transition matrix is then

$$P = \begin{array}{c} \\ 0 \\ 1 \\ 2 \\ 3 \\ 4 \end{array} \begin{array}{c} \begin{array}{ccccc} 0 & 1 & 2 & 3 & 4 \end{array} \\ \begin{pmatrix} 1 & 0 & 0 & 0 & 0 \\ 1/2 & 0 & 1/2 & 0 & 0 \\ 0 & 1/2 & 0 & 1/2 & 0 \\ 0 & 0 & 1/2 & 0 & 1/2 \\ 0 & 0 & 0 & 0 & 1 \end{pmatrix} \end{array} .$$

The states 1, 2, 3 are all nonabsorbing states, and from any of these it is possible to reach the absorbing states 0 and 4.

Hence the chain is an absorbing chain. Such a process is usually called a <u>random walk</u>. When a process reaches an absorbing state we shall say that it is absorbed.

THEOREM. In an absorbing Markov chain the probability that the process will be absorbed is 1.

We shall indicate only the basic idea of the proof of this theorem. From each nonabsorbing state, s_j, it is possible to reach an absorbing state. Let n_j be the minimum number of steps required to reach an absorbing state, starting from state s_j. Let p_j be the probability that, starting from state s_j, the process will not reach an absorbing state in n_j steps. Then $p_j < 1$. Let n be the largest of the n_j and let p be the largest of the p_j. The probability of not being absorbed in n steps is less than p, in 2n steps is less than p^2, etc. Since $p < 1$, these probabilities tend to zero.

For an absorbing Markov chain we consider three interesting questions: (a) What is the probability that the process will end up in a given absorbing state? (b) On the average, how long will it take for the process to be absorbed? (c) On the average, how many times will the process be in each nonabsorbing state? The answers to all these questions depend, in general, on the state from which the process starts.

Consider then, an arbitrary absorbing Markov chain. Let us renumber the states so that the absorbing states come first. If there are r absorbing states and s nonabsorbing states, the transition matrix will have the following canonical (or standard) form

$$P = \begin{array}{c} \\ r \\ s \end{array} \overset{\begin{array}{cc} r \text{ states} & s \text{ states} \end{array}}{\left(\begin{array}{c|c} I & 0 \\ \hline R & Q \end{array} \right)}.$$

Here I is an r-by-r identity matrix, 0 is an r-by-s zero matrix, R is an s-by-r matrix, and Q is an s-by-s matrix. The first r states are absorbing and the last s states are nonabsorbing.

In Section 1, we saw that the entries of the matrix P^n gave the probabilities of being in the various states, after n steps, starting from the various states. It is easy to show that P^n is of the form

$$P^{(n)} = \left(\begin{array}{c|c} I & 0 \\ \hline * & Q^n \end{array} \right)$$

where the asterisk * stands for the s-by-r matrix in the lower left-hand corner of P^n, which we do not compute here. The form of P^n shows that the entries of Q^n give the probabilities for being in each of the nonabsorbing states after n steps for each possible nonabsorbing starting state. (After zero steps the process must be in the same nonabsorbing state in which it started. Hence $Q^0 = I$.) By our first theorem, the probability of being in the nonabsorbing states after n steps approaches zero. Thus every entry of Q^n must approach zero as n approaches infinity, i.e., $Q^n \longrightarrow 0$.

Consider then the infinite series

$$I + Q + Q^2 + Q^3 + \ldots.$$

Suppose that Q were a nonnegative number x instead of a nonnegative matrix. To correspond to the fact that $Q^n \longrightarrow 0$ we take x to be less than 1. Then

$$1 + x + x^2 + \ldots = (1 - x)^{-1}.$$

It can be proved that the matrix series behaves in exactly the same way. That is,

$$I + Q + Q^2 + \ldots = (I - Q)^{-1}.$$

The matrix $N = (I - Q)^{-1}$ will be called the <u>fundamental matrix</u> for the given absorbing chain. It has the following important interpretation.

Consider the set of paths for the first n steps of an absorbing chain that starts in state s_i. Let X_k be a random variable which is 1 if the kth outcome is state s_j and 0 otherwise. Then

$$P(X_k = 1) = q_{ij}^{(k)}$$

and

$$P(X_k = 0) = 1 - q_{ij}^{(k)}$$

where $q_{ij}^{(k)}$ is the ijth entry of Q^k. This equation holds for k = 0 if we define $Q^0 = I$. Thus $E(X_k) = q_{ij}^{(k)}$. Let T_n be a random variable whose value is the total number of times in the first n steps that the chain is in state s_j. Then

$$T_n = X_0 + X_1 + \ldots + X_n$$

and

$$E(T_n) = E(X_0) + E(X_1) + \ldots + E(X_n)$$

$$= q_{ij}^{(0)} + q_{ij}^{(1)} + \ldots + q_{ij}^{(n)}$$

This expected value is the ijth entry of the series

$$Q^0 + Q^1 + \ldots + Q^n.$$

Let us define n_{ij} to be the limit of $E(T_n)$ as n tends to infinity. The quantity n_{ij} may be interpreted as "the expected number of times the chain is in s_j if it starts in s_i and proceeds until it is absorbed." We see from above that n_{ij} is the i,jth entry of the matrix

$$N = I + Q + Q^2 + \ldots,$$

that is, of the fundamental matrix. Thus we have answered question (c) as follows.

THEOREM. Let $N = (I - Q)^{-1}$ be the fundamental matrix for an absorbing chain. Then the entries of N give the expected number of times in each nonabsorbing state for each possible nonabsorbing starting state.

Example (Continued). In our example, the transition matrix in canonical form is

$$P = \begin{array}{c} \\ 0 \\ 4 \\ 1 \\ 2 \\ 3 \end{array} \begin{array}{c} \begin{array}{ccccc} 0 & 4 & 1 & 2 & 3 \end{array} \\ \begin{pmatrix} 1 & 0 & 0 & 0 & 0 \\ 0 & 1 & 0 & 0 & 0 \\ 1/2 & 0 & 0 & 1/2 & 0 \\ 0 & 0 & 1/2 & 0 & 1/2 \\ 0 & 1/2 & 0 & 1/2 & 0 \end{pmatrix} \end{array}.$$

From this we see that the matrix Q is

$$Q = \begin{pmatrix} 0 & 1/2 & 0 \\ 1/2 & 0 & 1/2 \\ 0 & 1/2 & 0 \end{pmatrix}$$

and

$$I - Q = \begin{pmatrix} 1 & -1/2 & 0 \\ -1/2 & 1 & -1/2 \\ 0 & -1/2 & 1 \end{pmatrix} .$$

Computing $(I - Q)^{-1}$, we find

$$N = (I - Q)^{-1} = \begin{array}{c} \\ 1 \\ 2 \\ 3 \end{array} \begin{array}{c} \begin{array}{ccc} 1 & 2 & 3 \end{array} \\ \begin{pmatrix} 3/2 & 1 & 1/2 \\ 1 & 2 & 1 \\ 1/2 & 1 & 3/2 \end{pmatrix} \end{array}.$$

Thus, starting at state 2, the expected number of times in state 1 before absorption is 1, in state 2 it is 2, and in state 3 it is 1.

We next answer question (b). That is, starting in a

nonabsorbing state, how long will it take on the average before reaching an absorbing state? If we add all the entries in a row, we will have the expected number of times in any of the non-absorbing states for a given starting state, that is, the expected time required before being absorbed. This may be described as follows:

THEOREM. Consider an absorbing Markov chain with s nonabsorbing states. Let C be an s-component column vector with all entries 1. Then the vector NC has as components the expected number of steps before being absorbed for each possible nonabsorbing starting state.

THEOREM. Let b_{ij} be the probability that an absorbing chain will be absorbed in state s_j if it starts in the nonabsorbing state s_i. Let B be the matrix with entries b_{ij}. Then

$$B = NR,$$

where N is the fundamental matrix and R is as in the canonical form.

Proof. Let s_i be a nonabsorbing state and s_j be an absorbing state. If we compute b_{ij} in terms of the possibilities, on the outcome of the first step we have the equation

$$b_{ij} = p_{ij} + \sum_k p_{ik} b_{kj},$$

where the summation is carried out over all nonabsorbing states k. Writing this in matrix form gives

$$B = R + QB$$

$$(I - Q)B = R,$$

and hence

$$B = (I - Q)^{-1}R = NR.$$

Example (Continued). In the random walk example we found that

$$N = \begin{pmatrix} 3/2 & 1 & 1/2 \\ 1 & 2 & 1 \\ 1/2 & 1 & 3/2 \end{pmatrix}.$$

From the canonical form we find that

$$R = \begin{pmatrix} 1/2 & 0 \\ 0 & 0 \\ 0 & 1/2 \end{pmatrix}.$$

Hence

$$B = NR = \begin{pmatrix} 3/2 & 1 & 1/2 \\ 1 & 2 & 1 \\ 1/2 & 1 & 3/2 \end{pmatrix} \begin{pmatrix} 1/2 & 0 \\ 0 & 0 \\ 0 & 1/2 \end{pmatrix}$$

$$= \begin{array}{c} \\ 1 \\ 2 \\ 3 \end{array} \begin{array}{c} \begin{array}{cc} 0 & 4 \end{array} \\ \begin{pmatrix} 3/4 & 1/4 \\ 1/2 & 1/2 \\ 1/4 & 3/4 \end{pmatrix} \end{array}.$$

Thus, for instance, starting from s_1, there is probability 3/4 of absorption in s_0 and 1/4 for absorption in s_4. Let us summarize our results. We have shown that the answers to questions (a), (b), and (c) can all be given in terms of the fundamental matrix $N = (I - Q)^{-1}$. The matrix N itself gives us the expected number of times in each state before absorption depending upon the starting state. The column vector NC gives us the expected number of steps before absorption, depending upon the starting state. The matrix NR gives us the probability of absorption for each of the absorbing states, depending upon the starting state.

The fact that we have been able to obtain these three descriptive quantities in matrix form makes it very easy to write a computer program which determines these quantities for a given absorbing chain matrix. The program ACHAIN carries out these computations. We have run the program for the random walk example with four nonabsorbing states. This is equivalent to two players matching pennies if they have a total of 5 pennies and agree to quit when one player gets all of the pennies.

ACHAIN

```
100 READ N,M
110 REM  N = NUMBER OF ABSORBING STATES
120 REM  M = NUMBER OF TRANSIENT STATES
130
140 MAT READ Q(M,M)
150 MAT READ R(M,N)
160 MAT C = CON(M,1)
170 MAT U = IDN(M,M)
180 MAT V = U-Q
190 MAT N = INV(V)
200 MAT T = N*C
210 MAT B = N*R
220 PRINT "MATRIX N ="
230 MAT PRINT N
240 PRINT
250 PRINT "MATRIX T ="
260 MAT PRINT T
270 PRINT
280 PRINT "MATRIX B ="
290 PRINT
300 MAT PRINT B
310 DATA 2,4
320 DATA 0,.5,0,0
330 DATA .5,0,.5,0
340 DATA 0,.5,0,.5
350 DATA 0,0,.5,0
360 DATA .5,0,0,0,0,0,0,.5
370 END
```

```
ACHAIN

MATRIX N =

1.6          1.0          .8          .4
1.6          1.0          .8          .4
 .8          1.6         2.4         1.2
 .4           .8         1.2         1.6

MATRIX T =

4
6
6
4

MATRIX B =

 .8           .2
 .6           .4
 .4           .6
 .2           .8
```

ACHAIN makes use of various matrix operations in BASIC to obtain the matrices N, T, and B, given matrices Q and R of an absorbing Markov chain. Matrix C (line 160) is defined to be the M X 1 matrix having all ones as its entries. (CON(R,S) is the RxS matrix having all ones as its entries.) Matrix U is defined in line 170 to be the M X M identity matrix. These matrices are then used to compute N, T, and B, where N is the fundamental matrix, T is a column vector giving expected number of steps before absorption depending on the starting state, and B is the matrix giving the probability of absorption for each absorbing state depending on the starting state.

EXERCISES

The examples referred to below are those in Section 1.

1. For what choices of a and b in Example 2 do we obtain an absorbing Markov chain?

   Ans. $a = 0$ or $b = 0$.

2. Show that Example 5 is an absorbing Markov chain.

3. Which of the genetics examples (Examples 7,8,and 9) are absorbing?

   Ans. Examples 8 and 9.

4. Find the fundamental matrix N for Example 8.

5. Verify that for Example 9 the following matrix is the inverse of $I - Q$ and hence the fundamental matrix N:

$$N = \begin{pmatrix} 8/3 & 1/6 & 4/3 & 2/3 \\ 4/3 & 4/3 & 8/3 & 4/3 \\ 4/3 & 1/3 & 8/3 & 4/3 \\ 2/3 & 1/6 & 4/3 & 8/3 \end{pmatrix}.$$

   Find NC and NR.

6. In the Land Of Oz example let us change the transition matrix by making R an absorbing state. This gives

$$\begin{array}{c} \\ R \\ N \\ S \end{array} \begin{array}{c} \begin{array}{ccc} R & N & S \end{array} \\ \begin{pmatrix} 1 & 0 & 0 \\ 1/2 & 0 & 1/2 \\ 1/4 & 1/4 & 1/2 \end{pmatrix} \end{array}.$$

   Find the fundamental matrix N, and also NC, and NR. What is

the interpretation of each quantity?

7. In Example 6, make states 0 and 3 into absorbing states. Find the fundamental matrix N, and also NC and NR, for the resulting absorbing chain. Interpret the results.

8. In the random walk example of the present section, assume that the probability of a step to the right is 2/3 and a step to the left is 1/3. Find N, NC, and NR. Compare these with the results of probability 1/2 for a step to the right and 1/2 to the left.

9. A number is chosen at random from the integers 1,2,3,4,5. If x is chosen, then another number is chosen from the set of integers less than or equal to x. This process is continued until the number 1 is chosen. Form a Markov chain by taking as states the largest number that can be chosen. What is the expected number of draws?

   Ans. 25/12.

10. Using the result of Exercise 9, make a conjecture for the form of the fundamental matrix if we start with integers from 1 to n. What would the expected number of draws be if we started with numbers from 1 to 10?

11. Three tanks fight a three-way duel. Tank A has probability 1/2 of destroying the tank it fires at. Tank B has probability 1/3 of destroying, and Tank C has probability 1/6 of destroying. The tanks fire together and each tank fires at the strongest opponent not yet destroyed. Form a

Markov chain by taking as states the tanks which survive any one round. Find N, NC, NR, and interpret your results.

12. Show that the following is an alternative method for finding the probability of absorption in a particular absorbing state, say s_j. Find the column vector d such that the jth component of d is 1, all other components corresponding to absorbing states are 0, and Pd = d. There is only one such vector. Component d_i is the probability of absorption in s_j if the process starts in s_i. Use this method to find the probability of absorption in state 1 in the random walk example given in this section.

13. The following is an alternative method for finding the expected number of steps to absorption: Let t_i be the expected number of steps to absorption starting at state s_i. This must be the same as taking one more step and then adding $p_{ij}t_j$ for every nonabsorbing state s_j. (a) Give reasons for the above claim that

$$t_i = 1 + \sum_k p_{ij}t_j,$$

where the summation is over the nonabsorbing states. (b) Solve for t for the random walk example. (c) Verify that the solution agrees with that found in the text.

Problems 14 and 15 problems concern the inheritance of color-blindness, which is a sex-linked characteristic. There is a pair of genes, g and G, of which the former tends to produce color-blindness, the latter normal vision. The G gene is dominant. But a man has only one gene, and if this is g, he is color-blind. A man inherits one of his mother's two genes, while a woman inherits one gene from each parent. Thus a man may be of type G or g, while a woman may be of type GG or Gg or gg. We will study a process of inbreeding similar to that of Example 7 of Section 1.

14. List the states of the chain. (Hint: There are six.). Compute the transition probabilities. Find the fundamental matrix N, and the vectors NC and NR.
15. Show that in both Example 9 of Section 1 and the example just given, the probability of absorption in a state having genes of a particular type is equal to the proportion of genes of that type in the starting state.

The remaining problems deal with the problem of gamblers ruin. A gambler starts with x dollars. He plays a game such that on each play he wins 1 dollar with probability p or loses 1 dollar with probability $q = 1 - p$. He must stop if his fortune drops to 0, ruin, or if it reaches a fixed number T (The bank is ruined!). We shall be interested in finding the probability w_x that, starting with x dollars, the gambler reaches T before 0. That is, that he is not ruined.

16. Show that the problem may be consider to be an absorbing Markov chain with state 0,1,2,..,T with 0 and T absorbing states. Use your simulation program to simulate a game in which the gambler has probability $p = .48$ of winning on each play, starts with 50 dollars and $T = 100$ dollars. Repeat this simulation 100 times and see how often the gambler is ruined.

17. For the general situation show that w_x satisfies the the following conditions:

    (a) $w_x = pw_{x+1} + qw_{x-1}$ for $x = 1,2,\ldots,T-1$

    (b) $w_0 = 0$

    (c) $w_T = 1$.

    Show that these conditions determine w_x. Finally, show that if $p = q = 1/2$ then

    $w_x = x/T$

    satisfies (a),(b) and (c) and hence is the solution. If $p \neq q$ show that

$$w_x = \frac{(q/p)^x - 1}{(q/p)^T - 1}$$

satisfies these conditions and hences gives the probability of the gambler winning.

18. Write a program to compute the probability w_x for given values of x,p,and T. Study the probability that in a game that is only slightly unfavorable, say p = .49 that the gambler will ruin the bank if the bank has significantly more money than the gambler.

3. ERGODIC MARKOV CHAINS.

A second important kind of Markov chain that we shall study in detail is a regular Markov chain defined as follows:

DEFINITION. A Markov chain is called a <u>regular</u> chain if some power of the transition matrix has only positive elements.

Of course, any transition matrix that has no zeros determines a regular Markov chain. The transition matrix of the Land of Oz example had $p_{NN} = 0$ but the second power P^2 has no zeros, so this too is a regular Markov chain. An example of a nonregular Markov chain is the chain with transition matrix

$$P = \begin{pmatrix} 1 & 0 \\ 1/2 & 1/2 \end{pmatrix}.$$

All powers of P will have a 0 in the upper right-hand corner.

The probabilistic interpretation of regularity is the following: A Markov chain is regular if there is some time at which it is possible to be in any of the states regardless of the starting state.

We shall now discuss two important theorems relating to regular chains. The proofs will be given in the following section.

THEOREM 1. If P is a transition matrix for a regular chain, then:

(1) The powers P^n approach a matrix W (that is, each entry of P^n approaches the corresponding entry of W).

(2) Each row of W is the same probability vector w.

(3) The components of w are all positive.

Recall that the ijth entry of P^n, $p_{ij}^{(n)}$, is the probability that the process will be in state s_j after n steps if it starts in state s_i. Thus (1) states that long-range predictions (i.e., large n), are approximately the same for all n. That is, $p_{ij}^{(n)}$ is approximately w_{ij} for all large n. Result (2) states that these long-range predictions do not depend upon the starting state. That is, $w_{ij} = w_j$ depends only on state s_j being considered and not on the starting state s_i. Thus the probability of being in s_j in the long run is approximately w_j, independently of the starting state.

Example. Recall that for the Land of Oz example, the eighth power of the transition matrix P was to three decimal places

$$P = \begin{array}{c} \\ R \\ N \\ S \end{array} \begin{array}{c} \begin{array}{ccc} R & N & S \end{array} \\ \begin{pmatrix} .4 & .2 & .4 \\ .4 & .2 & .4 \\ .4 & .2 & .4 \end{pmatrix} \end{array}.$$

Thus, to this degree of accuracy, the probability of rain eight days after a rainy day is the same as the probability of rain eight days after a nice day, or eight days after a snowy day. Theorem 1 only predicts that for large n the rows of P will become constant. It is interesting that this occurs so soon in our example. The next theorem gives a method for determing the limiting matrix W.

THEOREM 2. If P is a transition matrix for a regular chain, then there is a unique vector w which has components which add to one and is such that wP = w. This vector has positive components. $P^n \longrightarrow W$ where each row of W is the vector w.

A vector x with the property xP = x will be called a <u>fixed vector</u> for P. Thus the common row of W is the unique vector w which is both a fixed vector for P and a probability vector.

<u>Example</u> (Continued). By Theorem 1 we can find the limiting vector w for the land of Oz from the fact that

$$w_1 + w_2 + w_3 = 1$$

and

$$(w_1\ w_2\ w_3) \begin{pmatrix} 1/2 & 1/4 & 1/4 \\ 1/2 & 0 & 1/2 \\ 1/4 & 1/4 & 1/2 \end{pmatrix} = (w_1\ w_2\ w_3).$$

These relations lead to the following four equations in three unknowns.

$$
\begin{aligned}
w_1 + w_2 + w_3 &= 1 \\
1/2w_1 + 1/2w_2 + 1/4w_3 &= w_1 \\
1/4w_1 + 1/4w_3 &= w_2 \\
1/4w_1 + 1/2w_2 + 1/2w_3 &= w_3.
\end{aligned}
$$

Our theorem guarantees that these equations have a unique solution. If the equations are solved we obtain the solution:

$$w = (.4 \ .2 \ .4)$$

in agreement with that predicted from P^8.

Many interesting results concerning regular Markov chains depend only on the fact that the chain has a unique fixed probability vector that is positive. This property holds for a wider class of Markov chains.

DEFINITION. A Markov chain is called an <u>ergodic</u> <u>chain</u> if it is possible to go from every state to every other state.

Obviously a regular chain is ergodic, because if the nth power of the transition matrix is positive, then it is possible to go from every state to every other state in n steps. On the other hand, an ergodic chain may not be regular. For instance, if from a given state we can go to certain states only in an even number of steps and to others only in an odd number of steps, then all powers of the transition matrix will have 0's.

Example 1. An example of an ergodic Markov chain which is not regular is provided by Example 6 of section 1. Recall that the transition matrix for this example was

$$P = \begin{array}{c} \\ 0 \\ 1 \\ 2 \\ 3 \end{array} \begin{array}{c} \begin{array}{cccc} 0 & 1 & 2 & 3 \end{array} \\ \begin{pmatrix} 0 & 1 & 0 & 0 \\ 1/3 & 0 & 2/3 & 0 \\ 0 & 2/3 & 0 & 1/3 \\ 0 & 0 & 1 & 0 \end{pmatrix} \end{array} .$$

In this example, if we start in 0 we will on all even numbers of steps be in either state 0 or 2 and on odd numbered steps we will be in states 1 or 3.

For ergodic chains the fixed probability vector still exists, is strictly positive and unique, but we have to make slightly different interpetations. The following two theorems, which we will not be able to prove, furnish an interpretation for this fixed vector.

THEOREM 3. Let P be the transition matrix for an ergodic chain. Let A_n be the matrix defined by

$$A^{(n)} = \frac{I + P + P^2 + \ldots + P^{n-1}}{n}.$$

Then $A_n \longrightarrow W$ where each row of W is the unique fixed probability vector $w = wP$ for P

To interpret this theorem, let us assume that we have an ergodic chain that starts in state i. Let $X_m = 1$ if the mth step is to state j and 0 otherwise. The average number of times in state j in the first n steps is given by

$$h_j^{(n)} = \frac{X_0 + X_1 + X_2 + \dots + X_{n-1}}{n} .$$

But X_m takes on the value 1 with probability $p_{ij}^{(m)}$ and 0 otherwise. Thus $E(X_m) = p_{ij}^{(m)}$ and the ijth entry of $A^{(n)}$ give the expected value of $h_j^{(n)}$. That is, the expected proportion of times in state j in the first n steps when the chain starts in state i.

If we call being in state j "success" and any other state "failure" we could ask if a theorem analogous to the law of large numbers for independent trials holds. The answer is yes and is given by the following theorem.

THEOREM 4. Let $h_j^{(n)}$ be the proportion of times in n steps that an ergodic chain is in state s_j. Then for any $e > 0$,

$$P(|h_j^{(n)} - w_j| > e) \longrightarrow 0$$

independent of the starting state i.

We have observed that every regular Markov chain is also an ergodic chain. Hence Theorems 3 and 4 apply also for regular chains. For example, this gives us a new interpretation for the fixed vector $w = (.4\ .2\ \ .4)$ in the Land of Oz example. It predicts that in the long run it will rain 40 percent of the time in the Land of Oz, be nice 20 percent of the time, and snow 40 percent of the time. We illustrate this by writing a program to simulate the behavior of a Markov chain. SIMULATE is such a program.

```
SIMULATE

10 DEF FNF(X) = INT(1000*X+.5)/1000
100 READ N,T,A
110 MAT READ P(N,N)
120 MAT READ A$(N)
130 LET I = A
140 FOR S = 1 TO T
150    LET X = 0
160    LET U = RND
170    FOR J = 1 TO N
180       LET X = X + P(I,J)
190       IF U < X THEN 210
200    NEXT J
210    LET I = J
220    PRINT A$(J);
230    LET H(J) = H(J)+1
240 NEXT S
250 PRINT
260 PRINT
265 PRINT "WEATHER","NO. OF DAYS","PROPORTION OF DAYS"
266 PRINT
270 FOR I = 1 TO N
280    PRINT    A$(I),H(I),FNF(H(I)/T)
290 NEXT I
300 DATA 3,365,1
310 DATA .5,.25,.25
320 DATA .5,0,.5
330 DATA .25,.25,.5
340 DATA R,N,S
350 END
```

We have run the program for the Land of Oz example starting in state R.

SIMULATE

```
RRSSNSNRSSSRRRNRNRRRRSRNRSRNRSRRRRSSRSSNRRRSNSRRNRRRRNRNSNSRNS
NSSSSNSNSSRSSNSNSSSSRNRNRNSSSSRSNSSNRSSSNRRRNSSSSSSSSSNSRSRRRN
RSSSNSSNSSRRRSNRSNSNRRSNRNRNSSSSNSRRRNSRRNSSRSRSSRRNRNSNRNRNSS
SNSSRRSRRNSRRNRNRNSSNRSSNRRRRRRNSSNSSNRSRSSSSSRNSSSNSNSSRRSSSS
RRRRRNSNRSSNRSSSNRNRSSSNRRNRNRNSNSNRSNSSSSSSSSRSNSSNSSSNSRNRNRR
SSSSSNRRNRRRRRSNSNRRNSNRRRNRRNRNSSSSSRRRSSRSSRNRRRNSRRR
```

WEATHER	NO. OF DAYS	PROPORTION OF DAYS
R	130	.356
N	84	.23
S	151	.414

We note that the simulation gives a proportion of times in each of the states not too different from the long run predictions of .4,.2,.4 assured by Theorem 3.

The computation of the fixed vector w may be very difficult if the transition matrix is very large. It is sometimes possible to guess the fixed vector on purely intuitive grounds. Here is a simple example to illustrate this kind of situation.

Example 2. A white rat is put into the maze of Figure 2.

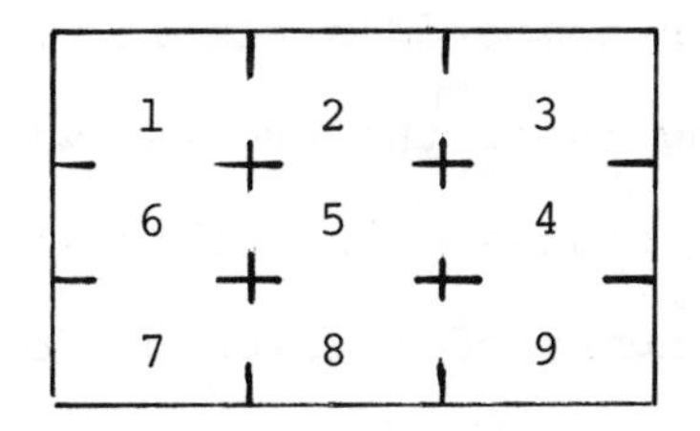

Figure 2.

There are nine compartments with connections between the compartments as indicated. The rat moves through the compartments at random. That is, if there are k ways to leave a compartment, it chooses each of these with equal probability. We can represent the travels of the rat by a Markov chain process with transition matrix given by:

$$P = \begin{array}{c} \\ 1 \\ 2 \\ 3 \\ 4 \\ 5 \\ 6 \\ 7 \\ 8 \\ 9 \end{array} \begin{array}{c} \begin{array}{ccccccccc} 1 & 2 & 3 & 4 & 5 & 6 & 7 & 8 & 9 \end{array} \\ \begin{bmatrix} 0 & 1/2 & 0 & 0 & 0 & 1/2 & 0 & 0 & 0 \\ 1/3 & 0 & 1/3 & 0 & 1/3 & 0 & 0 & 0 & 0 \\ 0 & 1/2 & 0 & 1/2 & 0 & 0 & 0 & 0 & 0 \\ 0 & 0 & 1/3 & 0 & 1/3 & 0 & 0 & 0 & 1/3 \\ 0 & 1/4 & 0 & 1/4 & 0 & 1/4 & 0 & 1/4 & 0 \\ 1/3 & 0 & 0 & 0 & 1/3 & 0 & 1/3 & 0 & 0 \\ 0 & 0 & 0 & 0 & 0 & 1/2 & 0 & 1/2 & 0 \\ 0 & 0 & 0 & 0 & 1/3 & 0 & 1/3 & 0 & 1/3 \\ 0 & 0 & 0 & 1/2 & 0 & 0 & 0 & 1/2 & 0 \end{bmatrix} \end{array} .$$

That this chain is not regular can be seen as follows: from an odd numbered state the process can go only to an even numbered state, and from an even numbered state it can go only to an odd number. Hence, starting in s_1 the process will be alternately in even and odd numbered states. Therefore each power of P will have 0's for the odd numbered entries or for the even numbered entries in row 1. On the other hand, a glance at the maze shows that it is possible to go from every state to every other state, so that the chain is ergodic.

To find the fixed probability vector for this matrix we would have to solve ten equations in nine unknowns. However, it would seem reasonable that the times spent in each compartment, in the long run, should be proportional to the number of entries to each compartment. Thus we try the vector with jth component the number of entries to the jth compartment. This is

$$x = (2\ 3\ 2\ 3\ 4\ 3\ 2\ 3\ 2).$$

It is easy to check that this vector is indeed a fixed vector so that the unique probability vector is this vector normalized to have sum 1. This is,

$$w = (1/12\ 1/8\ 1/12\ 1/8\ 1/6\ 1/8\ 1/12\ 1/8\ 1/12).$$

As we know $w = (w_1,\ldots,w_r)$ is the unique solution to

$$w = wP$$

and

$$\sum_i w_i = 1.$$

We can use this fact to develop a method for computing w which is easy to program. We do this as follows: Let

$$R = (I - P).$$

Let $\bar{R}$ be the matrix R with the last column replaced by a column of all one's. Then the vector w that we are looking for satisfies

$$(1) \quad w\bar{R} = (0,0,\ldots,1).$$

On the other hand assume that $\bar{w}$ is a vector which satisfies (1). The equations determined by (1) state that for $i = 1,2,..,r-1$

$$(2) \quad \sum_k \bar{w}_k p_{ki} = \bar{w}_i$$

and

$$(3) \quad \sum_i \bar{w}_i = 1$$

From (2) and (3) we see that

$$\sum_k \bar{w}_k p_{kr} = \sum_k \bar{w}_k (1 - \sum_{j=1}^{r-1} p_{kj})$$

$$= 1 - \sum_{j=1}^{r-1} \bar{w}_j$$

$$= \bar{w}_r.$$

We have thus shown that $\bar{w}$ is a fixed vector for P and has components which sum to 1. This proves that $\bar{w}$ is the unique fixed vector which is also a probability vector for P. But this means that the equation (1) has a unique solution. Recall that a matrix A has an inverse if and only if the equation $xA = 0$ has only the zero solution. If $\bar{R}$ did not have an inverse we could add a solution of $x\bar{R} = 0$ to w to obtain a new solution to (1). But since we have shown that (1) has a unique solution, $\bar{R}$ has an inverse and we can find w by

$$w = (0,0,\ldots,1)\bar{R}^{-1}.$$

The program LIMIT uses this result to compute the fixed vector for the ergodic chain obtained by considering the gas diffusion model of Example 6 of Section 1 with the number of balls equal to 5. The transition matrix for this case is:

$$P = \begin{pmatrix} 0 & 1 & 0 & 0 & 0 & 0 \\ .2 & 0 & .8 & 0 & 0 & 0 \\ 0 & .4 & 0 & .6 & 0 & 0 \\ 0 & 0 & .6 & 0 & .4 & 0 \\ 0 & 0 & 0 & .8 & 0 & .2 \\ 0 & 0 & 0 & 0 & 1 & 0 \end{pmatrix}.$$

```
LIMIT

100 DEF FNF(X) = INT(1000*X+.5)/1000
110 DIM P(20,20),I(20,20),X(20,20),A(1,20)
120 READ S
130 MAT A = ZER(1,S)
140 MAT I = IDN(S,S)
150 MAT READ P(S,S)
160 MAT X = I - P
170 FOR I = 1 TO S
180    LET X(I,S) = 1
190 NEXT I
200 MAT X = INV(X)
210 PRINT "LIMITING VECTOR W ="
220 PRINT
230 FOR I = 1 TO S
240    PRINT X(S,I);
250 NEXT I
260 DATA 6
270 DATA 0,1,0,0,0,0
280 DATA .2,0,.8,0,0,0
290 DATA 0,.4,0,.6,0,0
300 DATA 0,0,.6,0,.4,0
310 DATA 0,0,0,.8,0,.2
320 DATA 0,0,0,0,1,0
330 END
```

```
LIMIT

LIMITING VECTOR W =

.03125  .15625  .3125  .3125  .15625  .03125
```

By theorem 4 we can interpret these values for w_i as the proportion of times in each of the states in the long run. For example, the proportion of times in state 0 is .03125 and the proportion of times in state 2 is .3125. The astute reader will guess that these numbers are actually the binomial distribution 1/16,5/16,10/16,10/16,5/16,1/16. We could have guessed this answer in the following manner. If we consider a particular molecule, it simply moves randomly back and forth between the two sides of the box. Thus suggests that the equilibrium state should be just as if we randomly distributed the 5 molecules in the two sides of the box. If we did this, the probability that there would be exactly j molecules in one side of the box would be given by the binomial distribution with $n = 5$ and $p = 1/2$.

EXERCISES

1. Which of the following matrices are transition matrices for regular Markov chains?

(a) $$P = \begin{pmatrix} .5 & .5 \\ .5 & .5 \end{pmatrix}$$

(b) $$P = \begin{pmatrix} .5 & .5 \\ 1 & 0 \end{pmatrix}$$

(c) $$P = \begin{pmatrix} 1 & 0 \\ 0 & 1 \end{pmatrix}$$

(d) $$P = \begin{pmatrix} 1/3 & 0 & 2/3 \\ 0 & 1 & 0 \\ 0 & 1/5 & 4/5 \end{pmatrix}$$

(e) $$P = \begin{pmatrix} 0 & 1 \\ 1 & 0 \end{pmatrix}$$

(f) $$P = \begin{pmatrix} 1/2 & 1/2 & 0 \\ 0 & 1/2 & 1/2 \\ 1/3 & 1/3 & 1/3 \end{pmatrix}$$

2. Show that the 2 x 2 matrix

$$P = \begin{pmatrix} 1-a & a \\ b & 1-b \end{pmatrix}$$

is a regular matrix if and only if neither a nor b is 0 and they are not both equal to 1.

3. Find the fixed vector for the matrix in Exercise 2 for each of the cases listed there.

Ans. $w = (b/(a+b) \quad a/(a+b))$

4. Find the fixed probability vector w for each of the following regular matrices.

(a) $P = \begin{pmatrix} 75 & .25 \\ .5 & .5 \end{pmatrix}$

(b) $P = \begin{pmatrix} .9 & .1 \\ .1 & .9 \end{pmatrix}$

(c) $P = \begin{pmatrix} 3/4 & 1/4 & 0 \\ 0 & 2/3 & 1/3 \\ 1/4 & 1/4 & 1/2 \end{pmatrix}$

5. Let the process start in s_1 and compute $w^{(1)}$, $w^{(2)}$, $w^{(3)}$ for the matrices in Exercises 4(a) and 4(b). Do they approach the fixed vectors of these matrices?

6. Consider the transition matrix in Exercise 2 with a = b = 1. Prove that this chain is ergodic but not regular. Find the fixed probability vector, and interpret it. Show that P^n does not tend to a limit.

7. Consider the transition matrix of Exercise 2 with a = 0 and b = 1/2. Compute the unique fixed probability vector, and use

your result to prove that the chain is not ergodic.

8. Show that the matrix

$$P = \begin{pmatrix} 1 & 0 & 0 \\ 1/4 & 1/2 & 1/4 \\ 0 & 0 & 1 \end{pmatrix}$$

has more than one fixed probability vector. Find the matrix that P^n approaches, and verify that it is not a matrix all of whose rows are the same.

9. Prove that, if a 3-by-3 transition matrix has the property that its column sums are 1, then (1/3 1/3 1/3) is a fixed probability vector for it. State a similar result for n-by-n transition matrices. Interpret these results for ergodic chains.

Exercises 10 - 15 refer to the examples given in Section 1.

10. Show that Examples 8 and 9 are not ergodic chains.

11. Show that, for reasonable assumptions concerning a and b, Example 2 is a regular chain. (Hint: Use the results of Exercise 2.)

12. Using the result of Exercise 3, give the approximate number of people who are told that the President will run in Example 2. Interpret the fact that this proportion is independent of the starting state.

13. When an independent trials process is considered to be a Markov chain, what is its fixed probability vector? Is the

chain always regular? Illustrate this for Example 3.

14. Show that Example 1 is an ergodic chain, but not a regular chain. Find its fixed probability vector w. Show that w_j is the probability that box 1 would contain exactly j balls if we were to remove the balls and place them randomly in the boxes. That is, we put each ball, one at a time, in the boxes with probability 1/2 for each choice of the box.

15. Show that Example 2 is regular and find the limiting vector w.

16. A factory has two machines only one of which is used at any given time. A machine breaks down on any given day with probability p. There is a single repairman who takes two days to repair a machine and can work on only one machine at a time. We form a Markov chain by taking as states the pairs (x,y), where x is the number of machines in operating condition at the end of a day and y is 1 if a day's work has been put in on a machine not yet repaired and 0 otherwise. The transition matrix is

$$\begin{array}{c} \\ (2,0) \\ (1,0) \\ (1,1) \\ (0,1) \end{array} \begin{array}{c} \begin{array}{cccc} (2,0) & (1,0) & (1,1) & (0,1) \end{array} \\ \begin{pmatrix} q & p & 0 & 0 \\ 0 & 0 & q & p \\ q & p & 0 & 0 \\ 0 & 1 & 0 & 0 \end{pmatrix} \end{array},$$

where $p + q = 1$. Prove the chain regular and find the fixed vector.

Ans. $(1/(p^2+1))(q^2 \quad p \quad qp \quad p^2)$

4.* PROOF OF THE LIMIT THEOREMS FOR REGULAR CHAINS

We shall prove here the fundamental results concerning regular Markov chains stated in the previous section. We shall first show that there are several equivalent ways to state our basic result. We shall then prove one, and hence all, of these equivalent results.

THEOREM 1. If P is an r-by-r transition matrix, the following statements are equivalent.

(a) $P^n \to W$ as n tends to infinity, where W is a matrix with each row the same w. The vector w is a probability vector.

(b) For any r component probability vector x

$$xP^n \to w$$

where w is a probability vector. The same vector w is obtained for all choices of x.

(c) For any r component column vector y,

$$P^n y \to c,$$

where c is a column vector with all components the same value. The particular value obtained depends upon the choice of y.

Proof. We first prove that (a) and (b) are equivalent. We must show that (a) implies (b) and that (b) implies (a). Assume first that (a) is true. Let x be any r-component probability vector. Since $P^n \to W$, we also have

$$xP^n \to xW.$$

By assumption W has all rows the same probability vector w. Hence

$$xW = w.$$

(see Exercise 12.) Thus

$$xP^n \to xW = w.$$

Thus (a) implies (b).

Assume now that (b) is true. Let x_i be a row vector with 1 in the ith component and 0 in all other components. Then $x_i P^n$ is the ith row of P^n. By (b) this approaches the same probability vector w for any choice of i. Hence all rows of P^n approach the same probability vector. This proves that (b) implies (a). We have thus shown that (a) and (b) are equivalent. A similar proof shows that (a) and (c) are equivalent. (See Exercise 8.)

The statement (c) has a simple interpretation as an averaging process. Let us consider the result of multiplying a column vector y on the left by a probability matrix P. Since the row sums of P are one, each component of the vector Py is an average of the components of the vector y. The components of P^2y are an average of the components of the vector Py. In general, the vector $P^n y$ is the result of n such averaging operations. It seems reasonable that this averaging process should, for suitable P, smooth out the differences that may originally have existed in the components of the first vector. We shall first prove a theorem which shows that this "smoothing" does take place and then use this result to prove (c). Since (a), (b), and (c) are equivalent, we will have established the validity of all three of these statements for regular chains.

THEOREM 2. Let P be an r-by-r transition matrix with no zero entries. Let d be the smallest entry of the matrix. Let y be a column vector with r components, the largest of which is M_o and the smallest m_o. Let M_1 and m_1 be the largest and smallest component, respectively, of the vector Py. Then

$$M_1 - m_1 \leq (1 - 2d)(M_o - m_o).$$

Proof. The component m_1 of Py is an average of the components of y. This average assigns weight at least d to the maximum component M_o of y. The result is surely greater than or equal to the average which would be obtained by assigning weight d to the component M_o and all the rest of the weight to the minimum component m_o That is,

$$(1) \quad m_1 \geq dM_o + (1-d)m_o.$$

The same argument shows that

$$(2)\ M_1 \leq dm_o + (1-d)M_o.$$

Multiplying (1) by -1, we have

$$(3)\ -m_1 \leq -dM_o - (1-d)m_o.$$

Adding (2) and (3), we have

$$\begin{aligned} M_1 - m_1 &\leq d(m_o - M_o) + (1-d)(M_o - m_o) \\ &= (1-2d)(M_o - m_o). \end{aligned}$$

This completes the proof.

THEOREM 3. Each of the statements (a), (b), and (c) in Theorem 1 is true for a regular transition matrix.

Proof. Since we proved in Theorem 1 that the statements (a), (b), and (c) of Theorem 1 are equivalent, it is sufficient that we prove any one of them. We shall prove (c). We prove it only for the case that P has no zero elements, and the extension to the general case is indicated in Exercise 7.

Let y be any r-component column vector. Let M_n and m_n be, respectively, the maximum and minimum components of the vector $P^n y$. The vector $P^n y$ is obtained from the vector $P^{n-1}y$ by multiplying on the left by the matrix P. Hence each component of $P^n y$ is an average of the components of $P^{n-1}y$. Thus

$$M_o \geq M_1 \geq M_2 \geq \cdots.$$

and

$$m_o \leq m_1 \leq m_2 \leq \ldots.$$

Each sequence is monotone and bounded

$$m_o \leq m_n \leq M_n \leq M_o.$$

Hence each of these sequences will have a limit as n tends to infinity.

Let M be the limit of M_n and let m be the limit of m_n. We shall prove that M = m. This will be the case if $M_n - m_n$ tends to 0. Let d be the smallest element of P. Then by Theorem 2,

$$M_n - m_n \leq (1-2d)(M_{n-1} - m_{n-1}).$$

From this we see that

$$M_n - m_n \leq (1-2d)^n(M_o - m_o).$$

Since d can be at most 1/2 (see Exercise 6), the difference $M_n - m_n$ tends to 0 as n tends to infinity. Every component of $P^n y$ lies between m_n and M_n. Hence each component must approach the same number c = M = m. This proves statement (c) of Theorem 1 and hence also statements (a) and (b) of Theorem 1.

It is clear from the proof of Theorem 1 that the vector w which appears in (a) and (b) is in fact the same vector. The following theorem identifies this vector as the unique fixed probability vector.

THEOREM 4. Let P, W, and w be as in Theorem 1 part (a). Then w is the unique fixed probability vector of P.

Proof: Since the powers of P approach W, $P^{n+1} = P^nP$ approaches W, but it also approaches WP; hence WP = W. Any one row of this matrix equation states that wP = w, hence w is a fixed probability vector. We must still show that it is unique. Let z be any probability vector which is a fixed vector of P. By Theorem 1 part (b), we know that zP^n approaches w. But since z is a fixed vector, $zP^n = z$. Hence z remains fixed but "approaches" w. This is possible only if z = w. Hence w is the only probability vector which is a fixed vector of P.

We note that Theorem 1(b) is stronger than needed for a Markov chain process. For such a process we are interested only in the various rows of P^n, and hence we need only the special cases in which x has 1 as one component and 0 as its other components. But we can consider a more general kind of process. Suppose that we introduce a 0th experiment, in which we decide by a random (chance) device what the starting state should be. If x is any probability vector, we may think of its components as giving the probabilities of starting in various states. Then xP^n will give the probabilities of being in various states after n steps. The theorem establishes the fact that, even in this more general process, the probability of being in s_j approaches w_j, independently of the starting state.

We also obtain a new interpretation for w. Suppose that our random device picks state s_i as a starting state with probability w_i, for all i. Then the probability of being in the various states after n steps is given by $wP^n = w$, and is the same on all steps. Hence this method of starting provides us with a process that is stationary (or in "equilibrium"). The fact that w is the only probability vector for which $wP = w$ shows that we must have a random device of exactly the kind described to obtain a stationary process.

EXERCISES

1. Define P and y by

$$P = \begin{pmatrix} .5 & .5 \\ .25 & .75 \end{pmatrix} \quad y = \begin{pmatrix} .5 \\ .5 \end{pmatrix}.$$

   Compute Py, P^2y, P^3y, and P^4y and show that the result, approach a constant vector. What is this vector?

2. Let P be a regular r x r transition matrix and let y be any r-component column vector. Show that the constant vector which $P^n y$ approaches is Wy. Here W is the limiting matrix for P^n.

3. Let P be a regular matrix and assume y is a vector such that $Py = y$. Show then that $P^n y = y$. Show that this means that y must be a constant vector, that is, a vector each of whose components is the same number. Hence the only fixed column vectors for a regular transition matrix are constant vectors. Prove that if y is any constant column vector then

$Py = y$.

4. Show that if

$$P = \begin{pmatrix} 1 & 0 & 0 \\ .25 & 0 & .75 \\ 0 & 0 & 1 \end{pmatrix} \quad y = \begin{pmatrix} 1 \\ .25 \\ 0 \end{pmatrix} \quad z = \begin{pmatrix} 0 \\ .75 \\ 1 \end{pmatrix}$$

then $Py = y$ and $Pz = z$. Why does this show that the Markov chain with P as transition matrix is not regular? (Hint: Use the result of Exercise 3.)

5. Describe the set of all fixed column vectors for the chain given in Exercise 4.

6. Let P be the transition matrix with positive entries and let d be the minimum entry in the matrix. Show that d is at most 1/2.

7. The theorem that $P^n \to W$ was proved only for the case that P has no zero entries. Fill in the details of the following extension to the case that P is regular. Because P is regular, for some N, P^N has no zeros. Thus the proof given shows that $M_{nN} - m_{nN}$ approaches 0 as n tends to infinity. However, the difference $M_n - m_n$ can never increase. Hence, if we know that the differences obtained by looking at every Nth time tend to 0, then the entire sequence must also approach 0.

8. Prove that parts (a) and (c) in Theorem 1 are equivalent.

9. Let P be a regular transition matrix and let y be a column vector which has a 1 in the jth component and 0 in all other

components. Prove that c, the limit of $P^n y$, is the constant column vector with each component equal to w_j.

10. Let P, y, and c be as in Exercise 9. Prove that the common component c_0 of c is not zero. (Hint: Let N be such that P^N has no zeros. Let m_N be the minimum component of $P^N y$. Show that $0 < m_N \leq c_o$.)

11. Use the results of Exercises 9 and 10 to prove that for a regular chain the limiting vector w has all positive components.

12. Show that if W is the limiting matrix for a regular chain then xW = w for any probability vector x.

13. Prove that in an r state ergodic chain it is possible to go from any state to any other state in at most r-1 steps.

14. Let P be the transition matrix of an r state ergodic chain. Prove that if the diagonal entries p_{ii} are positive, then the chain is regular. (Hint: Show first that if the process can go from s_i to s_j in k steps, it can also go in any larger number of steps. Then use Exercise 13 to prove that P^{r-1} is positive.)

15. Prove that if P is the transition matrix of an ergodic chain, then 1/2(I + P) is the transition matrix of a regular chain. (Hint: Use Exercise 14.)

16. Prove that P and 1/2(I + P) have the same fixed vectors.

17. Prove, using Exercises 15 and 16, that an ergodic chain has a unique fixed probability vector, and that this vector is

positive.

18. Prove, using Exercises 3, 15, and 16, that for an ergodic chain only the constant vector c satisfies the equation $Px = x$.

19. We are given a regular Markov chain with transition matrix P and with state $S = \{0,1,2,\ldots,r\}$. We form a new Markov chain called a <u>coupled chain</u> as follows: The states of the new chain are pairs (x,y) where x and y are elements of S. The coupled chain starts in a state (x,y) with $x \neq y$. The x and y components then proceed to change according to the transition probabilities determined by P independently until the first time a state of the form (x,x) is reached. From that time on the coupled chain moves only to states of this same form. The probability of moving from (x,x) to (y,y) is again determined by P. That is, we independently run two copies of the given chain until the two copies first arrive at the same state and then they stay together, moving as a single chain determined by P. Prove that the coupled chain will eventually reach a state of the form (x,x) and move as a single chain. How does this help explain the fact that the intial position of a regular chain has no long range effect?

20. Assume that you watched only the x coordinate of the coupled chain described in Exercise 19. What would be the nature of this process?

5. MEAN FIRST PASSAGE TIMES FOR AN ERGODIC CHAIN

In this section we consider two closely related descriptive quantities of interest for ergodic chains. Let P be the transition matrix of an ergodic chain with states $s_1, s_2, \ldots, s_r$. Let $w = (w_1, w_2, \ldots, w_r)$ be the unique probability vector such that $wP = w$. Then we know that in the long run the process will spend a fraction w_j of the time in state s_j. Thus if we start in a different state s_i it will, with probability one, reach state s_j. Another way to see this is the following: Let us change this chain into a new Markov chain by making one state, say s_j, into an absorbing state. We do this by replacing the jth row of P by a row with a 1 in the jth component and 0 otherwise. If we start at any state other than s_j this new process will behave exactly like the old one up to the first time that state s_j is reached. Since the original chain was an ergodic chain, it was possible to reach s_j from any other state. Hence the new chain will be an absorbing chain with a single absorbing state s_j. We know that for an absorbing chain the process will eventually reach an absorbing state no matter where it is started. Hence we know that, in the original chain, state s_j is sure to be reached, no matter what the starting state is. (By this we mean, of course, that the probability of reaching s_j approaches 1.)

Let N be the fundamental matrix for the new chain. The entries of N give the expected number of times in each state before absorption. In terms of the original chain these quantities represent the expected number of times in each of the states before reaching state s_j for the first time. The ith component of the vector Ne gives the expected number of steps before absorption in the new chain, starting in state s_i. Here e is a column vector with each component 1. In terms of the old chain this is the expected number of steps required to reach state s_j for the first time starting at state s_i.

DEFINITION. If an ergodic Markov chain is started in state s_i the expected number of steps to reach state s_j for the first time is called the <u>mean first passage time</u> from s_i to s_j. It is denoted by m_{ij}. By convention $m_{ii} = 0$.

<u>Example 1</u>. Let us return to the maze example of Section 3. We shall make this ergodic chain into an absorbing chain by making state 5 an absorbing state. For example, we might assume that food is placed in the center of the maze and once the rat finds the food, he stays to enjoy it. See Figure 3.

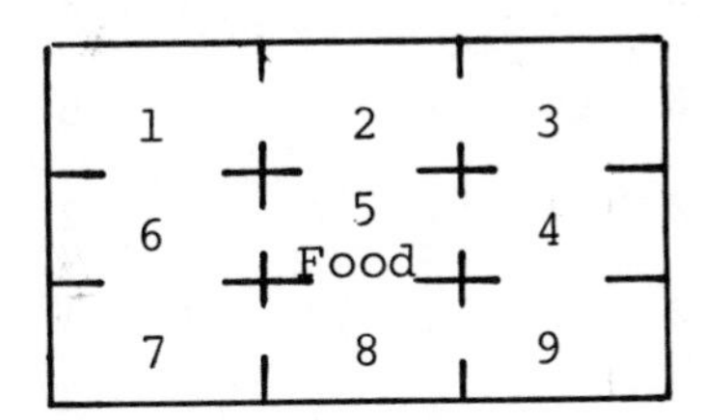

Figure 3.

The new transition matrix in canonical form is

$$
P = \begin{array}{c} \\ 5 \\ 1 \\ 2 \\ 3 \\ 4 \\ 6 \\ 7 \\ 8 \\ 9 \end{array}
\begin{array}{c|cccccccc}
5 & 1 & 2 & 3 & 4 & 6 & 7 & 8 & 9 \\
\hline
1 & 0 & 0 & 0 & 0 & 0 & 0 & 0 & 0 \\
\hline
0 & 0 & 1/2 & 0 & 0 & 1/2 & 0 & 0 & 0 \\
1/3 & 1/3 & 0 & 1/3 & 0 & 0 & 0 & 0 & 0 \\
0 & 0 & 1/2 & 0 & 1/2 & 0 & 0 & 0 & 0 \\
1/3 & 0 & 0 & 1/3 & 0 & 0 & 0 & 0 & 1/3 \\
1/3 & 1/3 & 0 & 0 & 0 & 0 & 1/3 & 0 & 0 \\
0 & 0 & 0 & 0 & 0 & 1/2 & 0 & 1/2 & 0 \\
1/3 & 0 & 0 & 0 & 0 & 0 & 1/3 & 0 & 1/3 \\
0 & 0 & 0 & 0 & 1/2 & 0 & 0 & 1/2 & 0
\end{array} .
$$

If we compute the fundamental matrix N we obtain

$$
N = 1/8 \begin{array}{cccccccc|c}
1 & 2 & 3 & 4 & 6 & 7 & 8 & 9 & \\
\hline
14 & 9 & 4 & 3 & 9 & 4 & 3 & 2 & 1 \\
6 & 14 & 6 & 4 & 4 & 2 & 2 & 2 & 2 \\
4 & 9 & 14 & 9 & 3 & 2 & 3 & 4 & 3 \\
2 & 4 & 6 & 14 & 2 & 2 & 4 & 6 & 4 \\
6 & 4 & 2 & 2 & 14 & 6 & 4 & 2 & 6 \\
4 & 3 & 2 & 3 & 9 & 14 & 9 & 4 & 7 \\
2 & 2 & 2 & 4 & 4 & 6 & 14 & 6 & 8 \\
2 & 3 & 4 & 9 & 3 & 4 & 9 & 14 & 9
\end{array} .
$$

The expected time to absorption for different starting states is given by the vector Ne,

$$Ne = \begin{matrix} 1 \\ 2 \\ 3 \\ 4 \\ 6 \\ 7 \\ 8 \\ 9 \end{matrix} \begin{bmatrix} 6 \\ 5 \\ 6 \\ 5 \\ 5 \\ 6 \\ 5 \\ 6 \end{bmatrix}.$$

We see that, starting from compartment 1, it will take on the average 6 steps to reach food. It is clear from symmetry that we should get the same answer for starting at state 3, 7, or 9. It is also clear that it should take one more step, starting at one of these states, than it would starting at 2, 4, 6, or 8. Some of the results obtained from N are not so obvious. For instance, we note that the expected number of times in the starting state is 14/8 regardless of the state in which we start.

A quantity that is closely related to the mean first passage time is the mean recurrence time which is defined as follows. Assume that we start in state s_i. Consider the length of time before we return to s_i for the first time. It is clear that we must return, since we either return on the first step or go to some other state and from any other state we know that we will eventually reach s_i.

DEFINITION. If an ergodic Markov chain is started in state s_i, the expected number of steps before the first return to s_i is called the mean recurrence time for s_i.

We shall need to develop some basic properties of the mean first passage times.

Consider the mean first passage time from s_i to s_j. Assume that $i \neq j$. This may be computed as follows: We consider the expected number of steps required, given the outcome of the first step, and multiply by the probability that this outcome occurs, and add. If the first step is to s_j, the expected number of steps required is 1. If it is to some other state s_k, the expected number of steps required is m_{kj} plus 1 for the step already taken. Thus

$$m_{ij} = p_{ij} + \sum_{k \neq j} p_{ik}(m_{kj} + 1)$$

or since

$$\sum_k p_{ik} = 1$$

$$(1) \quad m_{ij} = \sum_{k \neq j} p_{ik} m_{kj} + 1.$$

Similarly, starting in s_i it must take at least one step to return. Considering all possible first steps gives us

$$m_i = \sum_k p_{ik}(m_{ki} + 1)$$

$$(2) \quad m_i = \sum_k p_{ik} m_{ki} + 1.$$

Let us now define two matrices M and D. We define the ij entry m_{ij} of M to be the mean first passage time to go from s_i to s_j. M is called the <u>mean first passage matrix</u>. D will be the matrix with all entries zero except the diagonal entries d_{ii}. We define d_{ii} to be m_i. Then if E is an r x r matrix with all entries 1, we can write (1) and (2) in matrix form as

$$(3) \qquad M = PM + E - D$$

or

$$(4) \qquad (I-P)M = E - D.$$

We are now in a position to prove our first basic theorem.

THEOREM 1. For an ergodic Markov chain, the mean recurrence time for state s_i is $m_i = 1/w_i$, where w_i is the ith component of the fixed probability vector for the transition matrix.

Proof: Multiplying both sides of (4) by w, and using the fact that $w(I-P)=0$ gives

$$0 = wE - wD.$$

wE is a row vector with all entries 1. wD is a row vector with ith entry $w_i m_i$. Thus

$$(1,1,\dots,1) = (w_1 m_1, w_2 m_2, \dots, w_r m_r)$$

and

$$m_i = 1/w_i$$

as was to be proved.

Example 1 (Continued). In Section 3 we found that the fixed probability vector for our example was

$$w = (1/12 \quad 1/8 \quad 1/12 \quad 1/8 \quad 1/6 \quad 1/8 \quad 1/12 \quad 1/8 \quad 1/12).$$

Hence the mean recurrences are given by the reciprocals of these probabilities. That is,

$$(12 \quad 8 \quad 12 \quad 8 \quad 6 \quad 8 \quad 12 \quad 8 \quad 12).$$

Returning to the Land of Oz, we found that the weather in the Land of Oz could be represented by a Markov chain with states rain, nice, and snow. In Section 3 we found that the limiting vector was $w = (2/5 \quad 1/5 \quad 2/5)$. From this we see that the mean number of days between rainy days is 5/2, between nice days is 5, and between snowy days is 5/2. The above method for computing M is cumbersome and we give now a method better adapted to computing. We shall need a matrix which plays the role for ergodic chains that N does for absorbing chains. We shall use the matrix $Z = (I-P+W)^{-1}$. We must first show that this inverse exists and develop some of its properties.

THEOREM 2. Let P be the transition matrix of an ergodic Markov chain. Let W be the matrix with each row the fixed probability vector w for P. Then the matrix inverse

$$Z = (I-P+W)^{-1}$$

exists.

Proof: Recall that a matrix R has an inverse if and only if the only column vector x such that $Rx = 0$ is the zero vector. We shall prove that this is the case for $R = (I-P+W)$. Let x be a column vector such that

$$(I-P+W)x = 0.$$

Then multiplying this equation by w we have

$$w(I-P+W)x = wx = 0$$

Thus

$$(I-P+W)x = (I - P)x = 0.$$

But this means that $x = Px$ is a fixed column vector for P. This can only happen if x is a constant vector (see Exercise 18 of Section 4). But we also know that $wx = 0$ and w has strictly positive entries. Thus the constant must be 0 and $x = 0$ as was to be proven.

THEOREM 3. Let $Z = (I-P+W)^{-1}$. Then

$$wZ = w$$

and

$$Ze = e.$$

Proof: We shall prove that $wZ = w$. The proof that $Ze = e$ is similar and is left for an exercise (see Exercise 18). Since $w(I-P) = 0$ and $wW = w$,

$$w(I-P+W) = w.$$

Multiplying both sides by Z yields

$$w = wZ.$$

THEOREM 4. Let x and h be column vectors related by

$$(5) \quad (I - P)x = h.$$

Then

$$(6) \quad x = Zh + c$$

where c is a constant vector each of whose components is wx.

Proof: Adding Wx to both sides of (5) gives

$$(I-P+W)x = h + Wx.$$

Multiplying by Z gives

$$x = Zh + ZWx.$$

Since the columns of W are constant, it follows from $Ze = e$ that $ZW = W$. Thus

$$x = Zh + c$$

where each component of c is the constant wx.

THEOREM 5. The mean first passage matrix M is determined from Z by $m_{ij} = (z_{jj} - z_{ij})/w_j$.

Proof: Recall that (4) states that $(I-P)M = E-D$. Thus a column of M, $m_{\cdot j}$ satisfies an equation of the form (5), namely

$$(I-P)m_{\cdot j} = h$$

with h given by

$$h = \begin{bmatrix} 1 \\ 1 \\ . \\ . \\ . \\ 1 \\ 1 \end{bmatrix} - \begin{bmatrix} 0 \\ 0 \\ . \\ 1/w_j \\ . \\ 0 \\ 0 \end{bmatrix} .$$

By Theorem 3, $Ze = e$. Thus $(Zh)_i = 1 - z_{ij}/w_j$. Thus by Theorem 4

$$\{m_{.j}\} = 1 - Z_{.j}/w_j + c$$

where c is a constant vector. But we know that $m_{jj} = 0$. Thus each component of c has the value

$$c_o = z_{jj}/w_j - 1.$$

This gives us

$$m_{ij} = (Z_{jj} - Z_{ij})/w_j.$$

The program ERGODIC modifies LIMIT to include the calculation of the mean first passage matrix Z.

ERGODIC

```
20 DIM P(20,20),I(20,20),X(20,20),W(20,20),M(1,20)
30 DIM C(20,1),Z(20,20),N(20,20)
120 READ S
130 MAT W = ZER(1,S)
140 MAT M = ZER(1,S)
150 MAT N = ZER(S,S)
160 MAT READ P(S,S)
170 PRINT "TRANSITION MATRIX"
180 MAT PRINT P;
190 PRINT
200 MAT I = IDN(S,S)
220 MAT X = I-P
230 FOR Z = 1 TO S
240    LET X(Z,S)=1
250 NEXT Z
260 MAT X = INV(X)
270 FOR Z = 1 TO S
280    LET W(1,Z) = X(S,Z)
290    LET M(1,Z) = 1/W(1,Z)
300 NEXT Z
310 MAT I = I-P
320 MAT C = CON(S,1)
330 MAT Z = C*W
340 MAT Z = I+Z
350 MAT Z = INV(Z)
360 FOR I = 1 TO S
370    FOR J = 1 TO S
380       LET N(I,J) = (Z(J,J)-Z(I,J))/W(1,J)
390    NEXT J
400 NEXT I
410 PRINT "AVERAGE TIME IN STATE I"
415 PRINT
420 FOR I = 1 TO S
424    PRINT I;
430    PRINT W(1,I);
435    PRINT;
440 NEXT I
450 PRINT
460 PRINT "MEAN RECURRENCE TIME"
465 PRINT
470 FOR I = 1 TO S
480    PRINT I;
490    PRINT M(1,I);
495    PRINT;
500 NEXT I
510 PRINT
520 PRINT "MEAN FIRST PASSAGE MATRIX"
```

```
530 MAT PRINT N;
540 DATA 6
550 DATA 0,1,0,0,0,0
560 DATA .2,0,.8,0,0,0
570 DATA 0,.4,0,.6,0,0
580 DATA 0,0,.6,0,.4,0
590 DATA 0,0,0,.8,0,.2
600 DATA 0,0,0,0,1,0
999 END
```

```
ERGODIC

TRANSITION MATRIX

   .000  1.000   .000   .000   .000   .000
   .200   .000   .800   .000   .000   .000
   .000   .400   .000   .600   .000   .000
   .000   .000   .600   .000   .400   .000
   .000   .000   .000   .800   .000   .200
   .000   .000   .000   .000  1.000   .000

AVERAGE TIME IN STATE I

 1   .031
 2   .156
 3   .313
 4   .312
 5   .156
 6   .031

MEAN RECURRENCE TIME

 1 32.000
 2  6.400
 3  3.200
 4  3.200
 5  6.400
 6 32.000

MEAN FIRST PASSAGE MATRIX

   .000  1.000  2.500  5.167 11.667 42.667
 31.000   .000  1.500  4.167 1 .667 41.667
 37.500  6.500   .000  2.667  9.167 4 .167
 4 .167  9.167  2.667   .000  6.500 37.500
 41.667 1 .667  4.167  1.500   .000 31.000
 42.667 11.667  5.167  2.500  1.000   .000
```

We have run the program for the Ehrenfest urn model with 5 balls. From the mean first passage matrix we see that the mean time to go from all the balls in urn 1 to 2 is 2.5 steps while the expected time to go from 2 balls in urn 1 to 0 balls in this urn is 37.5. This reflects the fact that there is a central tendency built into the model. Of course the physicist is interested in the gas of a large number of molecules, or balls, and so we must consider this example for cases that we cannot compute even with a computer.

Example 2. Let us consider the Ehrenfest model for gas diffusion for the general case of 2n balls. Every second one of the 2n balls is chosen at random and moved from the urn that it was in to the other urn. If there are i balls in the first urn then with probability $i/2n$ we take one of them out and put it in the second urn, and with probability equal to $(2n - i)/2n$ we take a ball from the second urn and put it in the first urn. We let the number i of balls in the first urn be the state of the system. Then from state i we can pass only to state $i - 1$ and $i + 1$, and the transition probabilities are given by:

$$p_{i,i-1} = i/2n$$
$$p_{i,i+1} = (2n-i)/2n$$
$$p_{ik} = 0, \text{ otherwise.}$$

This defines the transition matrix of an ergodic but not regular Markov chain (see Exercise 14, Sec. 3). Here the physicist is interested in long-term predictions about the state occupied. As previously indicated we can see the long range predictions as follows: After many repetitions of the above mixing process and knowing nothing about how the process started we would expect any one ball to have probability 1/2 of being in either urn. The probability then that we would have i balls in the first urn would be

$$w_i = \binom{2n}{i}/2^{2n}.$$

That is, the same as the distribution of the number of heads in 2n tosses of a fair coin. Thus the mean recurrence time for state i is

$$m_i = 2^{2n}/\binom{2n}{i} .$$

Consider in particular the central term $i = n$. We have seen that in 2n tosses of a fair coin the probability of exactly n heads is approximately $1/\sqrt{\pi n}$. Thus we may approximate m_n by $\sqrt{\pi n}$.

This model was used to explain a concept of reversibility in physical systems. Assume that we let our system run until it is in equilibrium. A graph is handed to you, and you are asked to tell if the outcomes were graphed in the natural order of time or with time reversed. It would seem that there should always be a tendency to move toward an equal proportion of balls so that the correct order of time should be the one with the most transition

Consider in particular the central term $i = n$. We have seen that in $2n$ tosses of a fair coin the probability of exactly n heads is approximately $1/\sqrt{\pi n}$. Thus we may approximate m_n by $\sqrt{\pi n}$.

This model was used to explain a concept of reversibility in physical systems. Assume that we let our system run until it is in equilibrium. A graph is handed to you, and you are asked to tell if the outcomes were graphed in the natural order of time or with time reversed. It would seem that there should always be a tendency to move toward an equal proportion of balls so that the correct order of time should be the one with the most transition from i to $i - 1$ if $i > n$ and i to $i + 1$ if $i < n$.

In Figure 4 we have given two simulations of this process for $n = 1000$. We have recorded the outcomes for 500 trials after

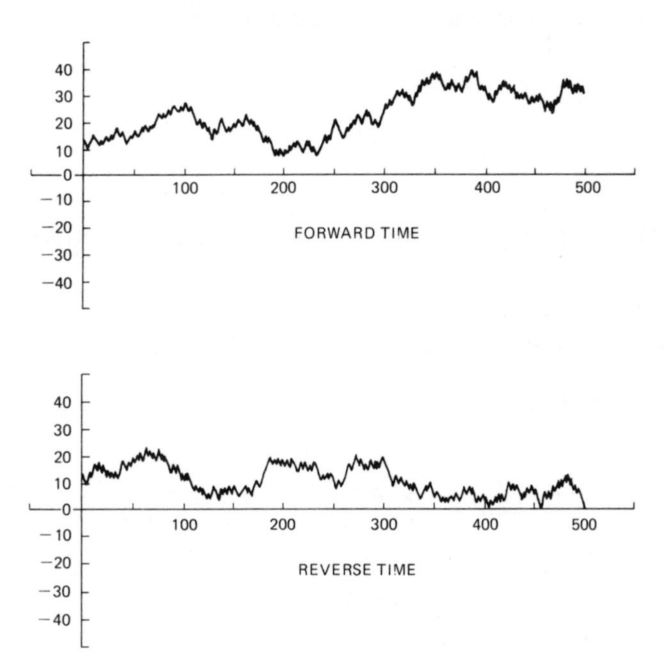

Figure 4.

the program has run 10000 trials. We have done this twice. The first time we have graphed the results in the correct order and in the second time with the order reversed.

There is no apparent time direction. The reason for this is that this process has a property called reversibility. Let us calculate for a general ergodic chain the reverse transition probability.

$$P(X_{n-1} = j \mid X_n = i) = \frac{P(X_{n-1} = j,\ X_n = i)}{P(X_n = i)}$$

$$= \frac{P(X_{n-1} = j)P(X_n = i \mid X_{n-1} = j)}{P(X_n = i)} = \frac{P(X_{n-1} = j)P_{ji}}{P(X_n = i)}$$

In general this will depend upon n since $P(X_n = i)$ and also $P(X_{n-1} = j)$ change with n. However, if we start with the vector w or wait until equilibrium is reached this will not be the case. Then we can define

$$p^*_{ij} = w_j p_{ji}/w_i$$

as a transition matrix for the process watched with time reversed.

Let us calculate a typical transition probability for the reverse chain $P^* = \{p^*_{ij}\}$. For example,

$$p^*_{i,i-1} = \frac{w_{i-1}p_{i-1,i}}{w_i} = \frac{C(2n,i-1)}{2^{2n}} \times \frac{2n-i+1}{2n} \times \frac{2^{2n}}{C(2n,i)}$$

$$= \frac{(2n)!}{(i-1)!(2n-i+1)!} \times \frac{(2n-i+1)i!(2n-i)!}{2n(2n)!} = \frac{i}{2n}$$

$$= p_{i,i-1}.$$

Similar calculations for the other transition probabilities show that $P^* = P$. When this occurs the process is called reversible. In the Ehrenfest model it shows that $w_i p_{i,i-1} = w_{i-1} p_{i-1,i}$. Thus in equilibrium the transition i,i-1 and i-1,i should occur with the same frequency. While many of the Markov chains that occur in applications are reversible, this is a very strong condition. For an example of a chain that is not reversible see Exercise 11.

EXERCISES

1. In Example 6 of Section 1 find for each state the mean first passage time to state 0. Do the same for state 2 and compare your results.
2. Consider Example 2 of Section 1 with a = .5 and b = .75. Assume that the President says that he will run. Find the expected length of time before the first time the answer is passed on incorrectly.
3. Find the mean recurrence time for each state in Example 6 of Section 1.

   Ans. (8 8/3 8/3 8).

4. Find the mean recurrence time for each state for Example 2 of Section 1 for a = .5 and b = .75. Do the same for general a and b.
5. A die is rolled repeatedly. Show by the results of this

section that the mean time between occurrences of any given number is 6.

6. For the Land of Oz example, make rain into an absorbing state and find the fundamental matrix. Interpret the result obtained from this chain in terms of the original chain.

7. For the Land of Oz example, write down the nine equations obtained by considering (1) for every i and j. Show then that the resulting equations have a unique solution.

8. In Example 1 of this section, assume that food is placed in compartment 5 and a trap is set in compartment 1. Form a Markov chain by making these states absorbing. Find the fundamental matrix N and the vectors Ne and NR. Interpret your results in terms of the original chain.

9. In Exercise 16 of Section 3, find the expected number of days between days when no machine is working. If both machines are in working condition at the end of a particular day, what is the expected number of days before the first day that no machine is working?

Ans. $(1+p^2)/p^2$ $(1+p)/p^2$.

10. Let X_m be the outcome for the mth draw in Example 2 of this section. If m is large we may assume

$$P(X_m = j) = \binom{2n}{j}/2^{2n}.$$

Find the expected value and variance of X_m.

11. Two players match pennies and have between them a total of 5

pennies. If at any time one player has all of the pennies, to keep the game going he gives one back to the other player and the game will continue. Show that this game can be formulated as an ergodic chain. Study this chain using the program ERGODIC.

12. Calculate the reverse transition matrix for the Land of Oz example. Is it reversible?

13. Show that every two state ergodic chain is reversible.

14. Give an example of a three state ergodic Markov chain that is not reversible.

15. Let P be the transition matrix of an ergodic Markov chain and P* the reverse transition matrix. Show that they have the same fixed probability vector w.

16. Show that any ergodic Markov chain with a symmetric transition matrix, i.e., $p_{ij} = p_{ji}$, is reversible.

17. Show that for an ergodic Markov chain Ze = e.

18. An ergodic Markov chain is started in equilibrium, i.e., with initial probability vector w. The *mean time in equilibrium to reach state* s_i is defined as

$$\bar{m}_j = \sum_k w_k m_{ki}.$$

Show that

$$\bar{m}_i = z_{ii}/w_i - 1.$$

19. Show that for an ergodic Markov chain

$$\sum_i m_{ri} w_i = \sum_j z_{jj} - 1 = K.$$

The constant K is called "Kemeny's constant." A prize is offered for the first person to give an intuitively plausible reason for the above sum to be independent of r.

BIBLIOGRAPHY

David, F. N.,"Games, Gods, and Gambling." New York,N.Y.: Hafner, 1962

Feller, W.,"An Introduction to Probability Theory and its Applications" Volume 1, 3rd Edition, New York: John Wiley, 1967.

Kemeny, J.G., and Kurtz, T.E.,"Basic Programming." 2nd Edition, New York: John Wiley, 1972.

Kemeny, J.G., and Snell, J.L., "Finite Markov Chains." Princeton, N.J.: Van Nostrand, 1960.

Kemeny, J.G., Snell, J.L., Thompson. "Finite Mathematics". 3rd Edition, Englewood Cliffs, N.J.: Prentice Hall, 1974.

Maistrov L. E. "Probability Theory, a Historical Sketch", New York N.Y.: Academic Press, 1974.

Ore, O. "Cardano, The Gambling Scholar", with a translation from Latin of Cardano's "Book of Games of Chance" by S. Gould., Princeton, N.J.: University Press, 1953.

Renyi, A., "Letters on Probability", translated by L. Vekerdi, Detroit, Mich.: Wayne State University Press, 1972.

Sprinchorn, E., "The Odds on Hamlet", The American Statistician, Vol.24. No. 5 (1964):14-17.